DIY Loudspeaker Building – 2nd edition

Packed with ideas on how to build your own speakers for home,

hi-fi or home theatre use

This work is copyright Julian Edgar © 2016.

First edition – May 2016

Second edition – June 2016

ALL RIGHTS RESERVED

Any unauthorised reprint or use of this material is prohibited. No part of this book may be reproduced or transmitted in any form or by any means, electronic or mechanical, including photocopying, recording, or by any information storage and retrieval system without express written permission from the author / publisher.

Edgar Media Pty Ltd

Contents

Building your own speakers - - - - -- 1

A quick course in subwoofer design - - - - 2

The Woofer Tester hardware / software package - - - 10

Building your own in-wall speakers - - - - 13

Building monster underfloor subwoofers - - - - 19

Home speakers on the ultra-cheap - - - - - 26

A complete speaker makeover - - - - - 29

An amplified compact speaker system - - - - 31

Good sound from old pipes! - - - - - 33

Superb spherical enclosure home hi-fi speakers - - - 37

Testing speakers - - - - - - - 44

Gaining quality loudspeaker drivers at low cost - - - 48

Adjusting tweeter levels - - - - - - 53

Using prebuilt enclosures to build a subwoofer - - - 56

Developing a compact speaker system - - - - 60

The author

Julian Edgar started his working life freelancing for photography magazines.

He then worked as a secondary school teacher for eight years, teaching senior humanities, before leaving teaching and becoming a full-time technical writer.

He edited a national Australian automotive print magazine before becoming editor of *AutoSpeed*, an online car magazine. He also wrote extensively for *Silicon Chip*, an electronics hobbyist magazine, while contributing articles to technical and automotive publications in Australia, the UK and the US.

He has been building and testing loudspeaker systems all his adult life, including in automotive and home settings.

Julian lives in a hamlet 80 kilometres north of Canberra, Australia. He spends much of the week playing in his home workshop – for the rest of the time, he works for a company providing training in high-level writing skills.

Other books by the author:

Automotive

21st Century Performance, Clockwork Media, 2000

High Performance Electronics for Cars, Silicon Chip Publications, 2004 (co-authored with John Clarke)

Amateur Car Aerodynamics Sourcebook, CreateSpace, 2013

DIY Car Electronic Modification Sourcebook, CreateSpace, 2013

DIY Testing of Car Modifications, CreateSpace, 2013

Tuning Programmable Engine Management, CreateSpace, 2014

Hybrid and Electric Cars Amateurs Sourcebook, CreateSpace, 2014

DIY Suspension Development, CreateSpace, 2015

Technical

Home Workshop Sourcebook, CreateSpace, 2013

Inventors and Amateur Engineers Sourcebook, CreateSpace, 2013

Using the Brilliant eLabtronics Modules, CreateSpace, 2015

Business and Government

Writing Effective Arguments, CreateSpace, 2015

Building your own speakers

It's fun and exciting building your own speakers – but there are also lots of ways of ending up with poor-sounding speakers that have taken a lot of time and money to put together.

This book therefore takes a different approach to that normally shown in speaker building books.

Firstly, few of the projects covered in this book use brand new drivers.

Why is this?

Well, if you buy new woofers, midranges, tweeters and crossovers, the sad reality is that you're already likely to be paying more than you would for similar quality, fully-built speakers purchased from a shop!

Instead, here I mostly use second-hand and salvaged drivers.

Secondly, for many of the projects, I use software and hardware to measure driver specs and then design enclosures to suit.

Specifically, I use Woofer Tester 2 to measure driver specs, both Woofer Tester 2 and BassBox to design enclosures, and Studio Six Digital FFT for analysing speaker response curves via an Apple i-device.

But doesn't having to buy these tools blow out of the water the cost savings achieved by using second-hand drivers?

No!

For example, Woofer Tester 2 is a one-off purchase that can be used to measure driver specs forever. And, as covered on page 10, using this tool allows you to work with pretty well any speaker drivers you can get. So you need find only *one pair* of good quality drivers salvaged from discarded speaker boxes or bought second-hand to save yourself the cost of Woofer Tester.

After that, it's all just a bonus!

(And another point. While you can of course buy drivers that come with Thiele-Small specs, and then use free on-line tools to design an enclosure to suit, in my experience the direct measurement of the driver specs with Woofer Tester gives far better results. Whether that's because of the variation in specs from driver to driver, or because specs are often measured incorrectly by the manufacturer or distributor, I am not sure.)

Thirdly, the projects covered in this book do not require high quality carpentry skills. This is an area very often brushed past in DIY speaker books – but it cannot be! To produce traditional speaker enclosures that can take pride of place in a lounge room or other home setting requires a *lot* of skilled work... so I don't take that approach.

Instead, speaker enclosures in this book are either hidden in the walls, ceiling or under the floor, or use pre-built items to easily form the enclosures.

Fourthly, I do not spend much time covering crossover design (or a lot on tweeter, midrange and woofer matching). What I suggest instead is that you source all these components from a prebuilt speaker, or car 'component' style parts are used.

For example, in the highest quality speaker design covered in this book (see page 37), I use 6.5 inch 'splits' that comprise the woofer, a dome tweeter and a dedicated full crossover. The crossover also includes a variable tweeter level control. Taking this approach removes a huge amount of 'hit and miss' in developing a crossover and matching the components. It's cheaper, too! (And how does that speaker sound? Superb!)

And fifthly – and this especially relevant if you're working on a tight budget – I also look at modifying existing fully-built speakers to give improved performance. This approach can cost almost nothing.

Finally, I make lots of use of free online tools – to design L-pads, to check crossover design values, and so on.

Building speakers that sound great and don't cost much is challenging, exciting, rewarding and fun.

A quick course in subwoofer design

- How speakers work
- Different enclosure designs
- Driver specifications
- Subwoofer software
- Designing a 12-inch car subwoofer

If there's one thing that everyone wants in a sound system, it's deep bass. The amount of the bass is the major criterion which many people use to judge the quality of sound - can it really shake you, or do you just hear the notes?

Starting this book with subwoofer design also introduces in an easy to understand way a lot of the ideas that we'll later be using.

How speakers work

Before we get into subwoofer enclosure design, you need to know something about how a speaker works.

A speaker operates by moving its cone back and forth. When the cone pushes forward it creates a high pressure wave which travels through the air, and when it moves back it causes a low pressure wave - sometimes called a 'rarefaction'. The ear picks up these air pressure waves and turns them into sounds.

The speed with which the cone moves will determine the pitch of the sound heard by the ear. If the cone is vibrating slowly - like back and forth say 50 times a second - then a deep note will be heard. Higher pitched sounds are made by the cone vibrating more quickly - 1000 times a second (expressed as 1000 Hertz) is a squeal, 8,000Hz is a screech, and 15,000Hz is higher than many people can hear at all.

Since we're interested here in bass, we'll stick to discussing the frequencies below 150Hz.

A woofer - which is a large speaker designed to work at low frequencies - will sound lousy if it is not mounted correctly. Foremost is the need to separate the front and rear of the speaker. To see why this is needed, imagine a bare woofer sitting on a bench. The speaker is being driven by a powerful amp, and is running music with a lot of bass. The cone will visibly move forward and backwards with the music, but the bass will be poor. The reason for this is that when the cone moves forward, a 'proper' pressure wave in the air isn't created. Instead, the air simply moves around the edges of the speaker frame, filling up the low pressure area created behind the cone. When the cone moves back again, the air flows the other way. Instead of pumping bass into the room, all the cone is doing is pumping air around its edges.

A bare subwoofer driver being tested on the bench will sound tinny. That's because the pressure waves can easily flow around the edges of the speaker frame, cancelling each other.

Separating the front and rear of the speaker doesn't mean that the driver must be in a sealed box - although that is one approach that works well. The separation of the pressure waves needs to occur acoustically, and in fact the enclosure design may well allow air to flow from the back of the speaker out the front through a port. But - and this is a key point - the enclosure needs to be designed so that the rear pressure waves add to the sound, not cancel it out.

But if "a speaker operates by moving its cone back and forth", why can't all speakers reproduce sound down to, say, 20Hz? After all, it's easy enough to drive a cone at this frequency. The answer is that the cone must actually *couple with the air*: it must impart energy to the air so that the waves travel through it. And achieving this requires very tricky design – and the trickiest bit is the subwoofer enclosure.

Different Designs

Four major types of speaker boxes are used. (Note that in each case it doesn't matter how many drivers are put into the box, whether they're facing in or out, etc.).

The designs are:

- **Sealed**

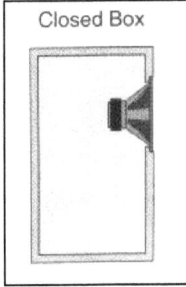

In this type of box, the sound waves coming from the back of the cone are effectively wasted. Instead of contributing to the sound that moves you, they're dissipating in the acrylic fluffy lining inside the sealed box. However, while this type of box produces less SPL (ie they are less efficient on a watts per dB basis) they are easy to make, have only a gradual bass drop-off as the notes get lower (and that drop-off can be counter-balanced by the rise in bass response that occurs within the closed confines of a room or car), and can potentially handle more power than a ported enclosure.

The effects of mismatching the speaker and its enclosure is also less severe when a sealed design is used, so if you have a driver of unknown specs, a sealed box is generally the only way to go. You can recognise a sealed box design easily – there are no openings in the box except for the driver(s).

- **Ported**

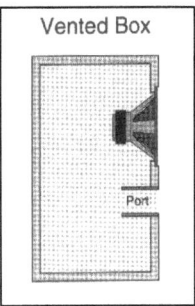

A briefly mentioned above, a ported enclosure additionally makes use of the energy coming from the back of the cone. It does this by using a connecting port (or vent) that joins the inside of the box with the outside. The port diameter and length are carefully sized so that the plug of air contained within the vent is excited into back and forth motion, but its movements are delayed just enough that when the cone of the driver is moving forward, so is the plug of air inside the port. In this way, the two air movements complement each other.

The advantages are twofold: firstly, the efficiency of the system is greater (ie more SPL per watt of amp power), and secondly the bottom-end bass response of the system can be improved over the use of a sealed box.

The downsides are that if the port isn't just right for the driver and box, boomy one-note bass can be the result, and even with well-designed ported boxes, at ultra-low frequencies the cone of the woofer becomes unloaded – which can cause it to be destroyed if you're not careful in how you set up the system.

And while the bass response holds on to lower notes, once it does start to fall away, it does so more quickly than with a sealed enclosure.

Ported designs can be easily recognised by the presence of one or more openings (usually round) that connect the other side of the driver to the atmosphere.

- **Passive Radiator**

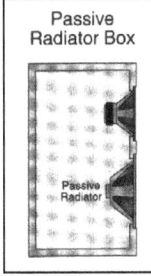

A passive radiator design is relatively rare – but it doesn't have to be. This type of design uses a passive radiator (a driver without a magnet) to act as a port – the cone and its suspension moves back and forth like the plug of air within a port, but at no times can the main driver become completely unloaded as is the case in a ported enclosure. The disadvantage is that passive radiators are not as easily bought – and finding the detailed specs on them which are needed to do good designs even rarer!

A passive radiator design looks to have two drivers in a sealed box (sometimes by just examining the system you won't be able to tell that one of the 'drivers' is in fact a passive radiator) although the radiator is usually bigger than the woofer.

- **Bandpass**

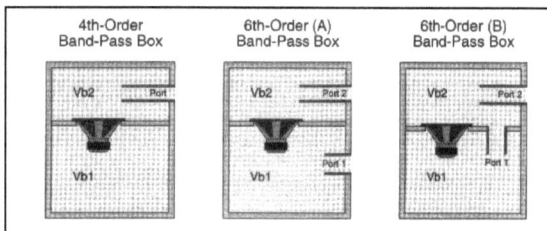

A bandpass design is a very tricky thing. Rather than producing frequencies from as high as the woofer can go – and then trailing them off at the other end as is determined by the enclosure design – a bandpass lets just a narrow spread of frequencies be emitted from the box.

Because it is producing only frequencies from (say) 30 – 90Hz, it can be more efficient that the other box designs – it's concentrating all of its energies just in this narrow field of frequencies. All the sound comes from the ports, with the driver itself buried inside.

There is a variety of bandpass designs – some mount the driver on the division between two boxes and vent just one volume, while others vent both boxes. Still others connect one box to the other by means of a vent and then have a further vent in the first box.

Bandpass boxes need very tricky design indeed, and if you're not careful, either the frequency spread (the range of bass notes produced by the sub) or the efficiency (how loud it is for a given input power) can drop right down.

In addition, the bandpass boxes are much more complex to make than other designs – there tends to be lots of pieces inside that need to be airtight, and fitting the ports inside the box can be a drama. However, get it right and you can have a loud and strong sub which is small and effective.

In a bandpass design the driver won't be visible (unless there's a plastic window fitted) – the speaker's cone does not connect directly with the air, except through a port.

Different Designs

For a subwoofer, key design criteria include:

- How loud do you want it?

- How much room do you have for it?

- What is the performance of the rest of the system? (Are there other drivers that will take the mid-bass load or does the sub have to do that as well?)

- Do you have a preferred driver? (It's on special at a bargain price this week...)

The design of subwoofers has been revolutionised by design software. Using it can help shape answers to a lot of the above questions – it's easy enough to find out, for example, that your chosen driver will need an enclosure that is too big for the available space, or that if you use a bandpass design you might need to beef-up the rest of the speakers to carry the response down to 100Hz.

Using subwoofer design packages

Above I covered the fundamentals of different sub box designs: sealed, ported, passive radiator and bandpass. Each has different characteristics in terms of the frequency response (the highest and lowest notes that can be reproduced), power handling, sensitivity (how loud the systems sounds for a given input power) and how large the enclosure needs to be.

In fact, there are so many variables that have an impact on these outcomes that back when dinosaurs were strutting the earth, most box design consisted of informed trial and error. Like, what's this sound like? Hopeless: try again… Is that better? Er, well not much…

The breakthrough came when two Australians - A.N.Thiele and Richard H. Small – realised that the behaviour of a speaker could be modelled in different enclosures if certain electrical and mechanical characteristics of the driver were known. These Thiele-Small characteristics, as they came to be known, are now available for all decent subwoofers. Strictly speaking, not all of the following are Thiele-Small parameters – but because most software design programs need them, I've included them here as well.

Commercial subwoofer design packages like BassBox are well worth the money.

Speaker Specs

- **Maximum Power (W)**

This shows the maximum input power that the speaker's voice coil can absorb without melt-down. (Note that it's got almost nothing to do with how loud the speaker will sound.)

Power is shown in watts, but there are all sorts of watts used. The only ones to take any notice of are watts RMS – other breeds of watts are sometimes overstated by as much as ten times…

- **Sensitivity (db @ 1w/1m)**

The amount of power being used to drive the speaker *and* its sensitivity *and* the sort of box it's mounted in will all determine the maximum sound pressure level. All three aspects are as important as the other - something almost always overlooked.

Speaker sensitivity is measured in dB at 1 watt/1 metre. In other words, it's how loud the speaker is when a microphone is placed a metre away from it and an input power of 1 watt is fed to the driver. Because the dB scale is used, small changes in this figure make a great deal of difference to how loud the speaker will be.

For each 1dB drop in speaker sensitivity, an amplifier 25 per cent more powerful will be needed to get the same SPL. Drop 3dB in sensitivity and you'll need an amp twice as powerful if you want the same sound level. Building a sensitive sub is an easy way of saving on (expensive) amp power.

Sub sensitivities range from 85 to 102dB, and - as a rule of thumb - bigger drivers are more sensitive than smaller cones. But remember that the box design is also a vital input into how loud the finished design will be.

- **Resonant Frequency (Fs)**

In general terms, this is the lowest frequency at which the speaker's cone can effectively couple itself with the air – there's only so much that fancy box designs can do. Subwoofers should have a resonant frequency of less than 50 Hz.

- **Vas**

The Vas figure is measured in either litres or cubic feet, and is an indirect measurement of how stiff the cone suspension is. The number actually indicates the volume of a closed box which would give the speaker the same stiffness as its suspension does.

A low Vas number means that the cone suspension is fairly stiff, while a high number shows that the cone suspension is more floppy.

- **Q – Qms, Qes, Qts**

In the same way that a car suspension needs to have its springs damped, so the moving mass of a speaker cone needs to be damped. 'Q' refers in fact to the opposite of damping – it's how much the resonance of the speaker is magnified.

Qms refers to the control exerted by the speaker's mechanical suspension – the spider and roll surround. Qes refers to the control exerted by the driver's electrical system – the voice coil and magnet. Qts is the total Q and is derived from both the Qms and Qes.

- **Xmax**

Xmax refers to the available travel of the cone. In other words, it's how far the cone can move before it hits the stops or the voice coil comes out of the core. A high Xmax figure means that the same size driver can move more air. Xmax can be measured in inches, cm or mm.

- **Others**

In addition to the above specs, you'll also find other variables like Re, Mms, BL, Sd and so on. Basically, the more data like this that is available, the better. You don't have to have an in-depth knowledge of what each of these parameters means – but you do have to have enough of them to plug into the program.

Designing

So you've found a driver that has available a whole heap of Thiele-Small parameters, you've got a good idea of the space available for the sub, and you want to start doing some virtual designs.

Now what?

The next step is to find a subwoofer design software package. If you do a web search you'll find a variety of free programs, or you can buy one. However, as you'd expect, the ones that you pay money for are a lot better than the freebies – better in terms of the data that can be generated, the different types of box designs that can be modelled, the 'help' support, and the accuracy of the results.

The software package that's used here as the example is BassBox, developed by US company Harris Technologies. In the 'Lite' version it's relatively cheap but can still do an excellent job.

For beginners, one of the real virtues of BassBox is that the provided help files represent a complete course in subwoofer design. The notes are to the point and very clear. In fact, this material alone is worth probably half of the cost of the software package.

The first step in the design process is to enter as many parameters are available for the driver. For example, let's imagine we are developing a car subwoofer that will fit into the available space, use a specific speaker driver, and have a response from 30-100Hz. It will be working with a 150W x 2 watt amp.

The available dimensions are 96cm wide with a usable height of 41cm. Up to 97cm depth is available – but the request was to keep the enclosure as small as possible. Looking at those dimensions, a possible volume of 157 litres is easily available – that's a lot. However, we'd expect to be able to do something reasonable much smaller than that.

The selected 12-inch driver has the following specs:

Impedance:	4 ohms
Power handling:	200 watts RMS
Fs:	25.6Hz
Qts:	0.409
Vas:	93.4 litres
Cone area:	490 square cm
Xmax:	10.5mm

With that data entered into BassBox, the first indication of the box type appears – a marker which is placed a little closer to 'closed box' than 'vented box' – indicating that this driver 'prefers' a sealed box design. However, this is not a hard and fast rule, and the fact that it's not right at one end of the continuum shows that there's some flexibility available.

Next, the program can be asked for its suggestions, under each box type. (But first the 'in-car' response needs to be turned on – this takes into account the dramatically increased bass that a sub can be generated in a car compared with the open air.)

Let's take a look at the screen displays as we work through the design process.

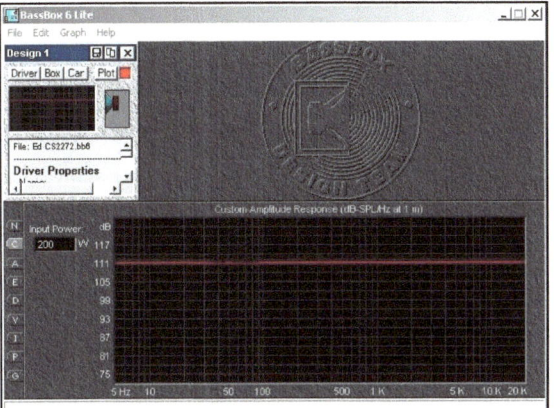

One button press later the sealed box volume is shown – just 31 litres when only small amounts of acrylic filling are used within the box! (And even smaller when the box is packed with acrylic filling, which oddly has the effect of increasing the apparent box size.) Another button press and the predicted in car response is shown (above) – it's ruler flat. This means that the predicted SPL stays the same at all frequencies – there's no major drop-off in bass.

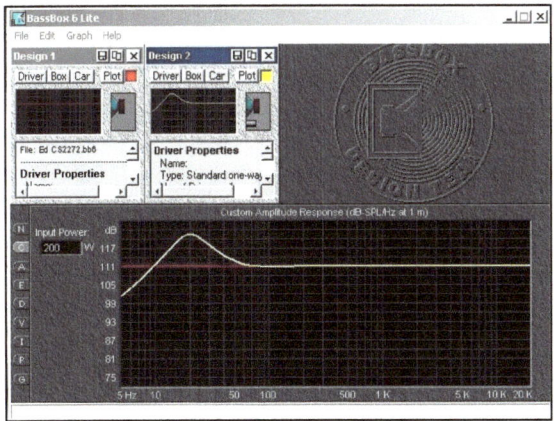

Next up, what about a suggested ported box? This time the program suggests that for max bass a massive box – 124 litres in size – should be used with a port 105mm in internal diameter and 370mm long. And the response? Boom, boom! The yellow line shows that at just 20Hz there is a major increase in SPL – it's up to 122dB, compared with the 111dB of the sealed design (both with an input power of 200W).

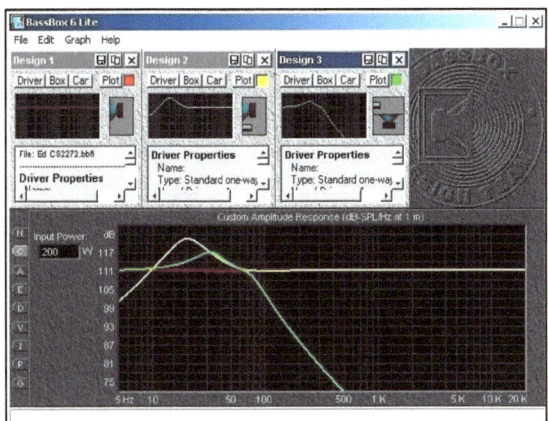

So what about a single-tuned bandpass design?

Press the buttons and it suggests a chamber volume of 30.6 litres and 26.4 litres but as can be seen by the green line, the result's not that great – there's a distinct peak at 31Hz. This is partly because the program's not taking into account the fact that the system is working in a car – it's easy enough to tweak this but straight off the suggestion list it's not good.

The suggested port size is also a whopper – 162mm diameter and 850mm long….

There are also other box designs that we could trial, but let's concentrate on these three.

While it first it looks like the decision's already made – after all, the sealed box was both smallest and gave the flattest response – but there's something else to keep in mind: how efficient the different enclosures are.

The frequency plots above have been made against SPL with an amp power of 200W – this gives the clearest picture of what is going on.

Remember, an increase in loudness by 3dB means you can halve the amp power needed – so if we could get the response of the ported and/or bandpass enclosures smoother but still staying loud, we'd have a winner. If the box is compact as well, anyway…

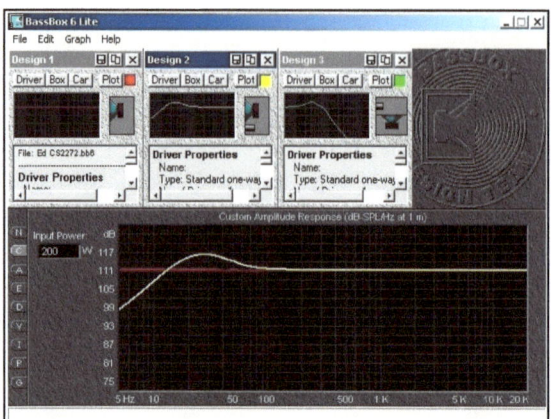
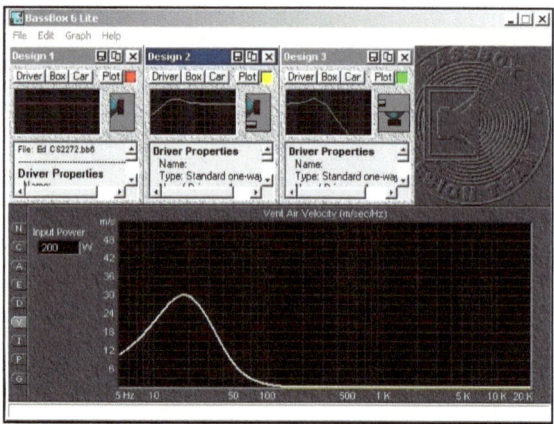

By reducing the ported box volume to 55 litres and using a port which is 75mm in diameter and 270mm long, we can get the response shown by the yellow line.

The red line (the sealed 31 litre box) has been kept on-screen as a comparison. The bass response of this modified ported design starts to rise smoothly from 100Hz and then peaks at 28Hz before smoothly dropping away again.

That will give smooth but strong bass.

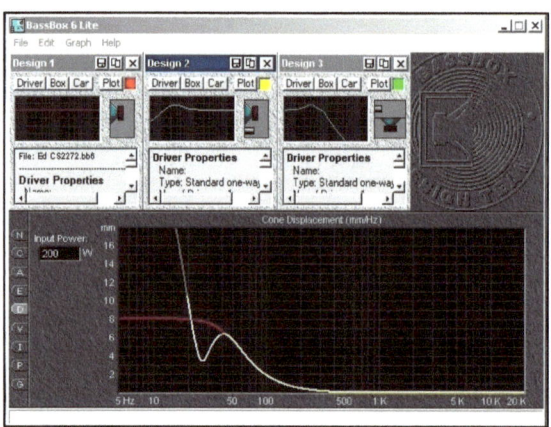

Another thing to check with this design is that the cone won't reach its Xmax – that it won't be bouncing off the bump stops. This graph shows both the sealed enclosure (red) and the revised ported enclosure (yellow). The change in density of the yellow line indicates that with an input power of 200W, the driver's Xmax will be reached at 20Hz, a very low frequency.

That's fine – if the figure had been 35Hz then there would have been real problems!

A final check is that the speed of the air movement inside the port won't be so high that whistles or buzzes are generated.

This graph shows that the max port speed will be 30 metres/sec – which again is fine.

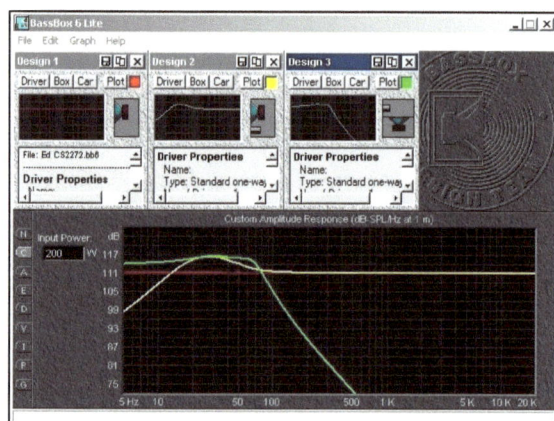

So that all looks great – but can we do better with a single tuned bandpass box? With one chamber at 50 litres and the other at 35 litres, and with the sound coming out through a single 110mm diameter, 171mm long port, the result is as shown here by the green line.

As can be seen, there's effectively no bass above about 75Hz but below that, it's strong and loud. But without other good quality speakers to carry the load down to 75Hz, there's going to be a mid-bass hole.

So to summarise, there are two major options:

- A sealed 31-litre box (red response)
- A ported 55 litre box with a 270mm long, 75mm vent (yellow response)

Note that if you go for the ported box you could realistically get away with a lot less amplifier power – perhaps half as much...

Of course there are plenty of other options – probably an infinite number of combinations, in fact. However, this brief overview shows the basics of the design approach that can be followed. As you can see, the opportunities to tailor the result to your preferences, budget and build skills are almost limitless.

Hints on enclosures

Ensure that ports are well sealed. Easiest is to do this sealing inside the enclosure.

The mouths of ports should not be blocked by internal stuffing.

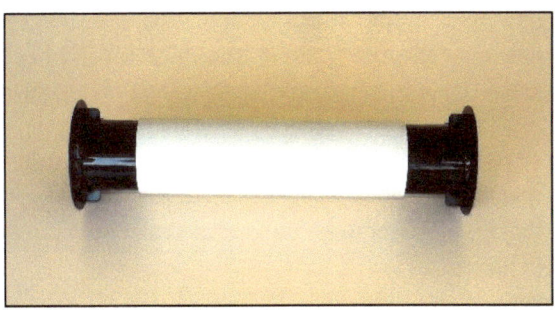

Port noise (or "chuffing") is more likely to occur if the ends of the port are simply cut off square. Bellmouths, as shown here, will reduce port noise. They can be bought as separate items or the plastic pipe can be heated and flared by pushing it down over the rear surface of a china bowl.

A pre-made speaker enclosure makes a good starting point. Here a hole is being drilled to allow a jigsaw to cut out the opening for a new port.

The Woofer Tester hardware / software package

- Off the shelf package
- Test almost any driver for Thiele-Small specs
- Allows effective use of second-hand and unknown drivers

As I've discussed, the Thiele-Small speaker parameters are especially important when designing woofers and subwoofers. In fact, without the Thiele Small (abbreviated to TS) specs of the driver, you're just guessing the box design – and the chances are overwhelming that your guess will be less than optimal!

So to design a good speaker enclosure, the TS specs are needed. Which is fine if you're buying a new driver.

But what if you've sourced a speaker that is literally an unknown? For example, a second-hand driver?

In that case, to get best results, the driver's Thiele Small specs need to be measured.

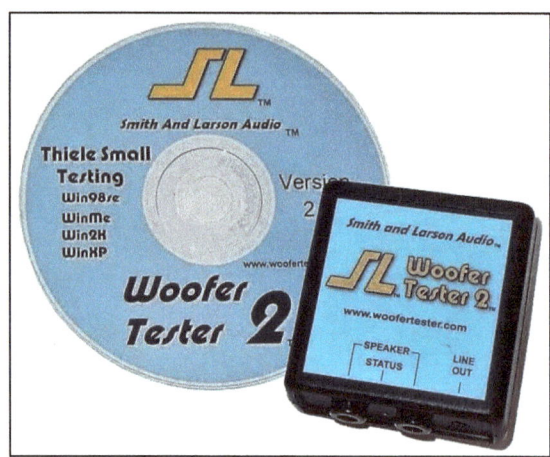

Woofer Tester 2 comprises a hardware and software solution for quickly and easily measuring the Thiele Small parameters of loudspeaker drivers. The software also allows you to design enclosures.

Woofer Tester 2

Woofer Tester 2 is a complete speaker test unit. This incredible piece of hardware plugs into the USB port of a PC or laptop and connects straight to the speaker under test. Open the software, press a button and within literally minutes many of the TS specs are measured. Do some more testing (eg by weighting the cone by a known amount) and the rest of the important specs are there in front of you – it's that easy!

Furthermore, the Woofer Tester 2 also includes a box design tool so you can also develop the enclosure without exiting the package.

And here's a really tricky thing - once the box is built, you can use the Woofer Tester 2 to test the speaker mounted in its enclosure, seeing if the measured performance matches the prediction. (Woofer Tester doesn't include a microphone, so you cannot directly measure frequency response – but indirectly you can get a good idea of what is happening, for example through the impedance plot.)

At US$160 at the time of writing, Woofer Tester 2 is not dirt cheap – but if you buy second-hand speakers, you need to use it only a few times to save that amount. (Of course, if you are using new drivers, you can also check to see if manufacturer's specs for the individual speaker is in fact correct – there is often some variation in the build from driver to driver.)

At its most complex, the Woofer Tester 2 has plenty of in-depth capability. But at its simplest (as I'll look at here), it's easy to get up and running.

Testing a driver

Woofer Tester 2 is used in the following manner. The software is installed and then the module calibrated using the provided test resistor. The provided alligator-clip leads are then used to connect the hardware to the speaker under test.

The speaker is placed on its back, ensuring that if it has a vented pole piece (ie vent hole in the magnet) this is not blocked.

In the software, the 'WT control' is opened from the View tab. The 'Q, Fs' test button is then pressed, and the test of the speaker starts automatically. The impedance (Re) of the speaker is measured and displayed, then the resonant frequency (Fs) and total Q (Qts) are ascertained. (Other factors are also measured, but I am trying to keep this simple!) This element of the test procedure takes a few minutes, during which you can see the impedance plot for the speaker developing on the screen in front of you.

So for example, the measured specs of a good 5-inch woofer might be:

- Re = 5.35 ohms
- Fs = 68.2Hz
- Qts = 0.48

Small weights are added to the cone (here coins have been used) and the change in resonant frequency of the driver is used by the software to calculate speaker compliance.

The next step is to press the 'Vas' test button. A dialog box pops up that asks you to add a weight to the upwards-facing cone. As I live in Australia, I use two Australian $2 coins that have a mass of 6.6 grams each. The added weight is, in this case, therefore 13.2 grams.

The effective diameter of the speaker cone is also manually measured (eg by callipers or a ruler) and then this figure is inputted.

The compliance (Vas) is then measured by the software, and the speaker sensitivity calculated. This takes about another 30 seconds.

This adds to our spec list for the 5-inch woofer:

- Vas = 11.9 litres
- Sens = 90.1 dB (at 1 watt, 1 metre)

Now to model a suitable enclosure.

Modelling enclosures

Opening the 'T/S Simulation' from the View tab gives you the in-built Thiele-Small speaker design software. The 'Test – Sim' button allows you to then import into the simulator the results you just got from the speaker test. Opening the 'Overlay' allows you to graph the predicted results of your various enclosure designs. These graphs change real time as you alter enclosure dimensions and port sizes. You can model vented, sealed, bandpass and passive radiator enclosure designs.

For example, using the tested specs described above for the 5-inch woofer, the software can be used to model an enclosure. In this case, a good result comes from using an 8-litre enclosure tuned to 57Hz using a port 25mm in internal diameter and 36.8mm long. The software can predict for this enclosure design the frequency response, impedance, phase – and lots of other parameters.

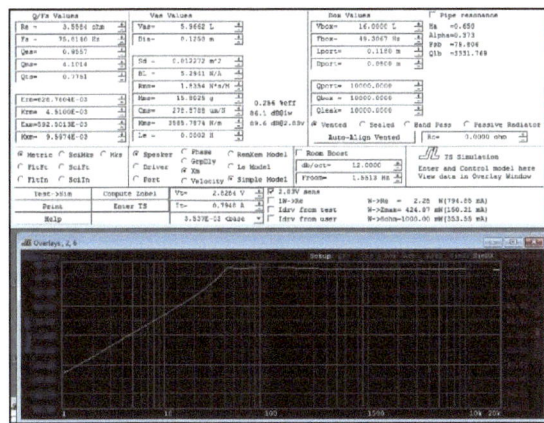

Here a speaker driver has been measured and the data imported into the enclosure design program that has been used to develop a 16 litre ported enclosure.

The next step is to build the suggested enclosure, then use the Woofer Tester 'Box test' function to test the completed speaker. You can then see how the tested box compares with the predicted performance, and make minor changes to the box design until the two line up perfectly.

Advantages and disadvantages

The enormous benefit of Woofer Tester 2 is that you can use it to measure the Thiele Small specs of drivers that are not provided with those specifications.

For example, very few – if any – 4-inch, 5-inch and 6-inch speakers come with Thiele-Small specifications. Thus, you have no real way of designing effective speaker enclosures for these small speakers.

As described earlier, you can also make excellent use of salvaged and second-hand speakers. Brand-name surround-sound home sound systems often have small, high quality drivers – but they're typically placed in poorly designed enclosures. Salvage some of these speakers, measure the driver's specs, model a good enclosure to suit, build the enclosure and then test it – and you can have quality home theatre sound at a fraction of the price you'd otherwise pay.

Measuring the specs for yourself?

If you do a search online, you'll find plenty of DIY techniques for measuring TS parameters.

You'll need a precision resistor, an AC multimeter that measures over a wide frequency range, and a frequency generator. And a lot of time spent doing very finicky measurements and plugging numbers into equations.

So it's certainly possible to do it manually – but it's time-consuming and likely to not be as accurate as using Woofer Tester 2.

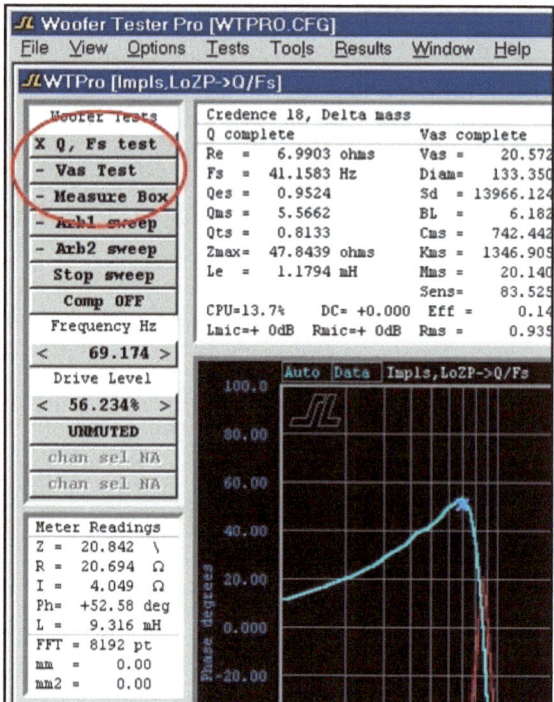

A variety of tests are available for ascertaining driver characteristics. Measuring specs such as Q, Fs and Vas are quickly and easily done.

And the disadvantages?

At US$160, Woofer Tester 2 is a not trivial in price. However, if you share it around among a few friends (and you each put in say $40 to buy it) then it will pay for itself very quickly.

Another disadvantage is that you need to already have some knowledge of speaker specs like Fs, Vas and Qts, and to have previously used some enclosure modelling software. I think that with no knowledge at all in these areas, the learning curve would be too great.

The provided handbook is OK (but not wonderful) and some of the software actions are not intuitive and are therefore a bit clunky.

However, overall, this is a must-have piece of equipment.

Building your own in-wall speakers

- High quality home sound without taking up any floor space
- Does not require Thiele-Small specs
- Allows effective use of second-hand speakers

Many different in-wall speakers are commercially available.

The vast majority use a 'bare' design: that is, the speakers are mounted on a baffle that is placed over a hole cut in the wall plasterboard. The space that is present behind the speakers therefore depends on the size of the wall cavity.

This volume can vary from wall to wall, so to cater for this, the designers of these speakers specify a fairly stiff speaker suspension on the mid / bass driver - otherwise, when installed over big spaces, the speaker would bottom-out when driven hard. In turn, this stiff suspension results in a high resonant frequency and high (and one-note) bass.

To cater for different in-wall volumes, most wall speakers use woofers with stiff suspension, giving a high resonant frequency and high Q. The result is poor bass.

A different approach is to use an in-wall enclosure. This type of speaker will sound much the same in different walls – a major advantage. This approach also allows the use of tuned (eg ported) enclosures. For example, Bose, with their 191 design, takes such an approach.

However, after testing the Bose 191, I decided to build my own enclosures and use the bass/mids, tweeters and crossovers from a pair of older Wharfedale Atlantic speakers I already had.

These Bose 191 in-wall speaker uses a ported enclosure that fits within the wall cavity. Using this approach, you can build your own system that not only sounds better than the Bose 191, but is also much cheaper.

Donor speakers

In standard form, each Wharfedale Atlantic uses three 8 inch drivers and a tweeter. The two lower drivers are housed in their own ported enclosure, and the mid/bass unit in a separate upper enclosure that is also ported. (The ports are on the back of the box.) The upper enclosure is about 15 litres in volume – a pretty good size for a custom-built in-wall enclosure.

I removed from the Wharfedales the mid/bass units, tweeters and crossovers and built them into a pair of new 15 litre enclosures, sized to fit in the walls.

The general approach, which is easily adopted, therefore consisted of:

- obtain a pair of compact, good-sounding speakers
- copy the enclosure design (volume, port length and size)
- use the drivers and crossovers in the new-shape, in-wall enclosures

This method has lots of benefits. It's cheap (much cheaper than buying the individual drivers and crossovers), requires no acoustic design, and allows (some) sound testing before tearing the walls apart!

The donor Wharfedale speakers. The upper driver, tweeter and crossover were used in each new wall speaker. In the original boxes, these are mounted within a separate, ported enclosure of 15 litres.

New boxes

One of the problems with building full enclosures into the walls (as opposed to using the wall cavities as the enclosures) is that you are severely limited in the thickness of box material you can use.

If the available wall depth is (typically) 100mm, then using thick particle board for the enclosure can easily reduce the available depth to just 70mm – too small to accommodate the depth of most mid/bass drivers.

To maximise the internal depth, I chose instead to use relatively thin 9mm medium density particle board.

To provide an internal volume of about 15 litres, the box dimensions are about 540 x 380 x 100mm. Note that the depth was dictated by the thickness of the wall framing (max 100mm) and the width by the distance between adjoining studs (400mm).

The boxes were assembled with butt joints, nailed and/or screwed into place. Pine cleats (40 x 20mm) were then placed at the corners of the box. Water clean-up Liquid Nails building adhesive was used on all joins.

A surplus of glue was used so that it was squeezed out of the joints as the panels were pushed together. A wet finger was then used to smooth this glue internally and externally along the seams, better sealing them. (The final panel, where internal access wasn't possible, used an even greater quantity of glue!)

The ports of the Wharfedale donor speakers were 50mm in internal diameter and 85mm long. However, a port of this size in the in-wall enclosures would put the internal end of the ports too close to the back wall of the boxes.

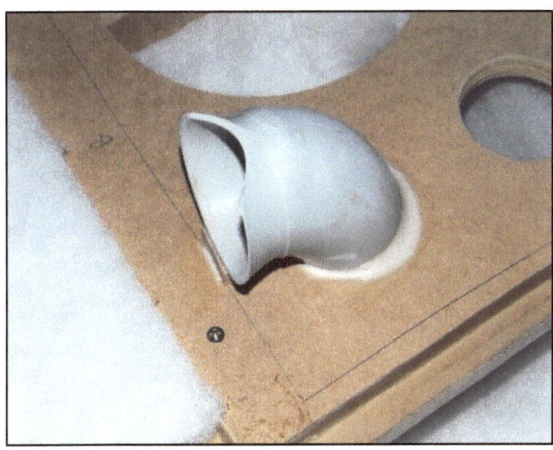

To achieve the required port clearance, I formed the ports from curved plumbing sections. The internal ends of the ports were bellmouthed by being heated and then forced down over a small ceramic bowl.

(Note that this design of plumbing has a smooth internal bend, not an inner radius that has a sharp edge at the change in direction. Many PVC pipe bends are like the latter.)

Part of the bell-mouth needed to be ground away to provide clearance to the back wall of the enclosure.

I wanted a paintable, professional looking metal grille and obtained it by buying the cheapest 8-inch in-wall speakers I could find on eBay.

When the cheap in-wall speakers arrived, I removed the drivers and crossovers, and then cut out the internal plastic panel with an electric jigsaw. (This needs very careful work.)

The result is a plastic frame into which the metal grille securely fits.

To give the grille frame clearance, I used a recessed front panel to mount the Wharfedale woofer, tweeter and the new port. The speaker panel mounts behind this front 'window' panel. Note also the strengthening cleats in the enclosure corners.

Internal surfaces of the enclosure were covered in a single layer of polyester quilt wadding, held in place with Liquid Nails adhesive. You can also see here the speaker mounting panel mounted behind the main baffle.

Testing

Prior to integrating into the walls, a frequency generator was used for initial testing of the speakers. I used a standalone instrument I already had, but if you don't have such an instrument, free and/or low cost downloads are available to turn a PC sound card or Apple i-device into a frequency generator.

The frequency generator should be connected to an amplifier and the amplifier used to drive the speaker under test. Be careful when driving speakers in this way: the power levels should be kept relatively low.

The frequency generator test showed that the new-shaped speaker had response that extended down to around 50Hz and was strong from about 70Hz to beyond the range of my hearing – excellent as a mid / bass and tweeter combination.

Working at this stage with just the first finished enclosure, I then connected the speaker to an amplifier and played music.

To be honest, I was surprised at how good it sounded – it was as good as the original Wharfedale (when of course the Wharfedale was played with just the mid/bass and tweeter connected). That it sounded much the same makes sense – but it felt quite odd when you looked at the weirdly-shaped enclosure.

Location, location!

I'd decided that the speakers should be mounted in the walls about midway between the floor and the ceiling. On paper, these positions worked well - the left-hand speaker located below a wall-mounted TV, and the right-hand one at a height to match. And that's where I cut the holes in the wall plasterboard.

The problem occurred when I placed the first wall speaker enclosure in the newly-cut wall hole – it sounded terrible! The sound was thin and lacked warmth - the upper bass / lower midrange had gone. (A test with the frequency generator showed a new and distinct hole around 100 - 150Hz)

It wasn't the enclosure or its drivers, because when I placed the system back on the floor it sounded fine again.

So what was it?

I had an assistant move the speaker box around while I listened. On the floor in the middle of the room - it sounded great. Up in the air, about a metre above the ground - fine. Over near the wall, still OK.

The assistant then slid the flat enclosure up the wall.

As soon as it reached the mid-point, half-way between the floor and ceiling, the sound was horrible! I then had my assistant raise the speaker still higher. When the grille reached a distance of about 10cm below the ceiling cornice, the sound was great.

So what could I do? Move the hole in the plasterboard... I extended the hole vertically by about 60cm, inserted the speaker enclosure - and everything was again fine.

Be careful where you mount your in-wall speakers!

Installation

Each wall speaker was installed using the following procedure.

The existing wall plasterboard was cut back so that the joins between old and new plasterboard would be located over timber studs (verticals) and noggins (horizontals).

More chipboard packing (arrowed) was used between the studs and the box to firmly locate the enclosure laterally. Liquid Nails glue was liberally applied underneath and on both sides of the box, effectively gluing the box in place between the studs. The front face of the box was located flush with the wall surface.

A new noggin was nailed between two of the existing studs - the enclosure was then placed on it, sitting on two packing pieces.

Plasterboard was cut to size fill the gaps and was glued in place using Gyprock Wet Area jointing compound (I have found this plastering adhesive to be particularly strong). Finishing plaster was then trowelled over all the joins and sanded smooth.

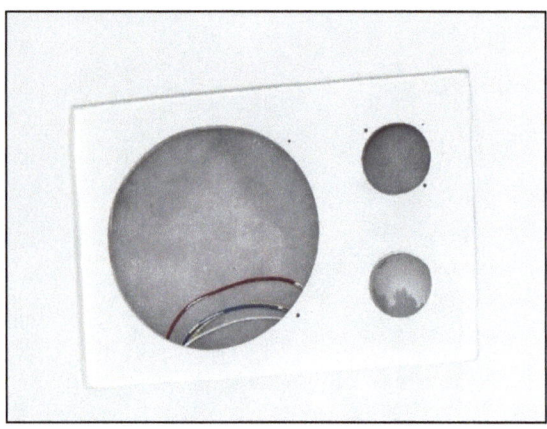

The inner baffle, the grille and the surround were painted and then….

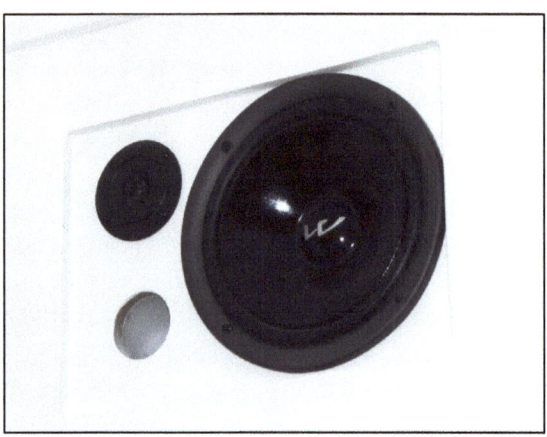

… the speakers were reinstalled.

Wrecking the walls?

If you have not done any wall plastering work before, you may be daunted by what is shown here. However, it's actually all pretty easy.

It certainly helps if you are already intending to repaint the room, but you don't have to be a master tradesperson to get a good result!

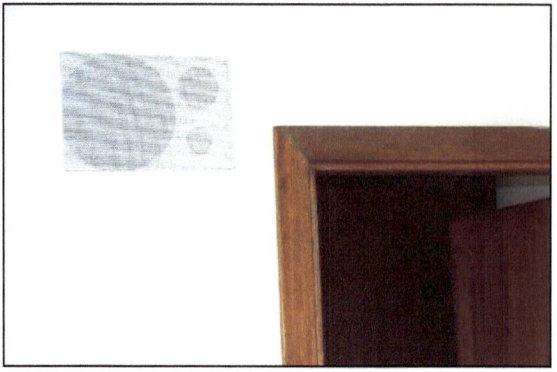

The grille and surround were glued into place and then the wall was painted. Note that the enclosure is much larger than the grille dimensions suggest.

Taking this installation approach gives a very secure final result – the speaker enclosure is part of the wall rather than just sitting in or on it.

(Footnote: And, two years later, no cracks have appeared – despite the speaker saving been worked hard.)

The sound

So how do they in-wall speakers sound? In one word - excellent. In fact, with the new enclosures integrated into the walls, a surprising discovery was made – *the drivers now sound better than they did in the original boxes*.

The bass from the in-wall speakers isn't particularly strong (that's what the underfloor woofers are for – they're covered next), although the bass would still be more than adequate for extension speakers, rear home theatre speakers or for general listening.

But it's the treble and midrange that are outstanding – the speakers having a 'transparency' of sound that is nuanced and very effective.

Building monster underfloor subwoofers

- Shake the house with these twin 15-inch subwoofers
- They fit under the floor in houses raised a little off the ground, or in the ceiling in other houses.

It began when we bought a house that needed renovating. The house also needed a big shed… so that's what it got first. But after the shed was built, and time passed, I decided I'd better get around to doing the work I'd long promised.

That's work like putting-up new plasterboard on the walls of the dining room and kitchen, replacing all the kitchen cupboards and bench-tops, painting, and laying ceramic floor tiles.

This, you must understand, is not work I find particularly interesting. It's OK – especially when you're learning new skills, but after only a short time it starts to lose its glamour. So, I thought while floor-tiling in the lounge room, what could I do to make this project a bit more interesting?

Hmm – how about built-in speakers? Like, subwoofers located in the crawl-space under the house, firing up through floor-mounted grilles? And then to complement them, speakers built into the walls? Walls speakers are not uncommon, but what was that about subwoofers?

In the past such ideas would have filled me with terror (what, tearing the floor and walls apart?) but these days, having become modestly proficient in flooring and plasterboarding, I reckoned it was all do-able.

Designs

My first thought was to mount a couple of drivers under the floor, with their enclosures being the whole of the atmosphere. That is, to use a genuine infinite baffle approach.

If you look around the Web you'll see some fantastic infinite baffle designs, where people have mounted drivers (often multiple, large drivers – like four 18 inch sub-woofers!) so that they work in under-floor or in-ceiling locations, pumping sound into the room through a short connecting duct.

The advantages of an infinite baffle design is that there is no speaker box to colour the sound, and with an infinite amount of air behind the cones, no stiffening of the driver's suspension through the trapped volume of springy air.

The disadvantages are that without that trapped air, cone excursions can be large and so you need a lot of driver area (ie lots of big speakers) so that you can use less power and have smaller cone excursions - and yet still get adequate air movement.

JBL GT5-15 drivers were used in the underfloor subwoofers. These are 15 inches in diameter and have a continuous power handling of 300 watts.

In my application, I could think of two further disadvantages of the infinite baffle approach.

Firstly, mounted under the floor, the rear of the speaker is exposed to outside air – and so, during rain and fog, airborne moisture. With the crawl space well-ventilated, I could see the moisture degrading the speakers over time.

Secondly, and for me this one was the big negative, the subwoofer cones are not particularly good heat insulators (why should they be?) and so in winter, I could see my heating bill being even bigger than usual! After all, it'd be much the equivalent of having two big holes in the floor...

So how could these disadvantages – especially the heat loss one – be overcome? The answer is to use an enclosure – to box-in the rear of the speaker. After all, there's plenty of room for a big box... or two.

But before going much further, I needed to bite the bullet and source the drivers that I was going to use in this underfloor location. I decided to use two drivers and started looking hard for suitable speakers.

I settled on some car subwoofers – the JBL GT5-15. These are 15 inches in diameter and have a continuous power handling of 300 watts.

With the speakers in-hand, and with full Thiele-Small parameters available, I could start computer modelling different enclosure designs. I've long used BassBox Lite so I fired it up and started modelling.

Models

The first design I modelled was the infinite baffle approach. Using the BassBox 'sealed box' program option (but with an inserted volume of 999999999 litres!) the software indicated a -3dB point of 33Hz. Pretty good – in the real world, as opposed to marketing literature, 33Hz is very low in note.

Using the software, I then trialled a sealed box – picking an arbitrary volume of 300 litres. The modelled response changed very little over the infinite baffle - after all, 300 litres is a pretty big box! In fact, when rounded, the -3dB point stayed at 33Hz.

But what about a ported box? The design of this is more complex, but by using a volume of 315 litres and a box tuned frequency of 20.5Hz (achieved in the software with a rectangular port 200 x 100 x 360mm long), the -3dB point could be stretched waaaaay down to 17.6Hz! That's another whole octave lower than with the other designs – at 17Hz you wouldn't be hearing the bass, you'll be feeling the house shake.

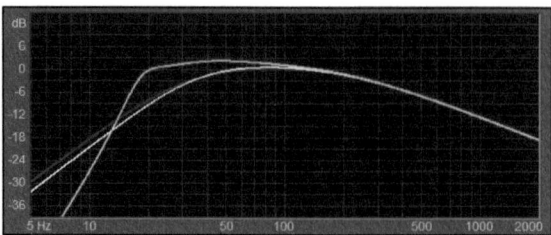

This graph shows the modelled response of the three designs. Red is the infinite baffle, yellow is the sealed box of 300 litres, and green is the ported box of 315 litres.

Note that these were never final designs – just indicative of what could be achieved with each approach. And the trial modelling showed that a ported box was best.

Logistics

The next step was to work out how it was all going to be achieved. One of the biggest stumbling blocks was to source grilles suitable for floor use. I wanted them to be able to be trodden on, to be appropriately sized and to, as much as possible, not look weirdly out of place in a lounge room. They'd be inset in a neutral coloured, ceramic tiled floor. (The floor I was now tiling!)

I started by looking at ducted heating vents (they call them 'registers') but these were generally smaller than I liked. In fact, the biggest I could find with an acceptable style was only 14 x 6 inches – just half the area of the driver.

I then contacted grille manufacturer Rayson Industries in Melbourne (Australia) and found the company very helpful. It turned out they had some metal floor grilles that they were discontinued items – they were therefore cheaper than usual. These were available in a variety of sizes - I bought two 254 x 356mm grilles in a beige colour. These cost about AUD$70 each. The grilles are powder-coated steel, are very strong and look decent.

Strong powder-coated steel mesh grilles were used for the floor openings.

Next, how were the boxes to be got under the house? There is just a single entrance area to the crawlspace – a door about 550mm square. It would be possible - but very difficult - to build the enclosures under the house (there's not even sitting-up room), so it would be best to make them outside and then insert them through the access door. That meant a maximum width/height of about 500mm.

So for a box volume of (say) 300 litres, the dimensions would be about 500 x 500 x 1200mm long. The overall speaker diameter was 390mm, and the required cut-out 365mm – so that worked OK with a face dimension of 500mm.

On paper that all seemed fine - but how would such an enclosure sit under the floor? What was the orientation and spacing of the floor joists? Observation and measurement showed that the spacing was too close (at 400mm) for the enclosure (and driver) to nestle between the joists. The box would therefore have to be positioned up against the underside of the joists, with an extension sealing the enclosure to the floor hole.

Hmm, now how to integrate the port?

Typically, the port of a vented enclosure exits the baffle at a place other than where the main speaker is located.

However, in my case, all the sound from each enclosure had to come through single 254 x 356mm floor grilles – there could not be a separate grille for the port. The vent would therefore need to work through the same floor opening as the driver.

So how to do that? I decided to have the port enter from the side, feeding air into the extension that connected the speaker to the floor grille.

Further design work resulted in this final configuration: 200 litres volume with two curved, right-angled 100mm diameter ports.

Final design

Speaker specs:

F_s = 27.3Hz

Q_{ms} = 5.7

V_{as} = 106.5 litres

X_{max} = 14.5mm

S_d = 855 sq cm

Q_{es} = 0.76

R_e = 3.63 ohms

L_e = 2.42mH

Z = 4.3 ohms

BL = 16.53Tm

P_e = 300 watts

Q_{ts} = 0.67

2.83 V SPL = 92dB

The calculated box properties were:

V_b = 200 litres

F_b = 24.6Hz

Q_l = 5

F_3 = 20.7Hz

Two vents

D_v = 100mm

L_v = 330mm

Materials and construction

Because the boxes would be to an extent unprotected from the weather (the crawl-space is well ventilated and so exposed to fog and wind-borne rain), and also because of the low cost, I chose to construct the boxes from particle board flooring.

I used 19mm thick, "green tongue" sheets. This material has a mass of 13.1kg / square metre and is available in a variety of sheet sizes. It cuts easily with either a coarse-bladed electric jigsaw or a circular saw.

The 19mm thick chipboard panels were cut with a circular saw. Note the clamped guide that gives straight cuts

The completed underfloor sub. The extensions nearest the camera fit up between the floor joists, with both the driver and the two ports connected to the room through a mesh grille set into the floor.

Butt joints (rather than mitres) were used; however, full-length cleats were added to strengthen every butt join. These cleats were made from 40 x 20mm pine. Every joint was both glued and screwed, with the screws connecting to the cleats rather than to the particle board.

Water clean-up Liquid Nails building adhesive was used. Lots of glue was placed at every joint so that it squeezed out as the panels were screwed together. A wet finger was then used to smooth this glue internally and externally along the seams, better sealing them.

The spacing of the floor joists meant that the main portion of the enclosure needed to butt against the underside of the joists, rather than sliding up between them. In turn this required that a narrower extension piece be used to connect the enclosure to the floor vent.

The side parts of the extension are formed in-situ by the floor joists, while the ends of the extension comprise pieces attached to the box.

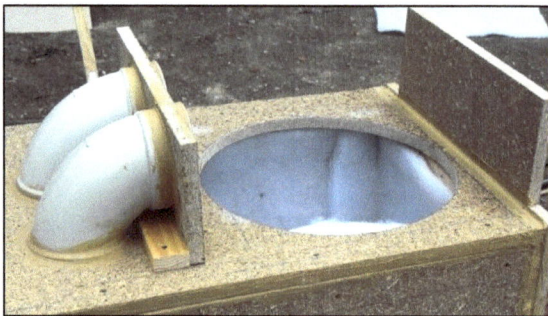

Two ports enter one of the extension pieces from the side. Each port is 100mm in diameter and 330mm long. The ports are formed from curved plumbing fittings with extension pieces inserted in each end.

Prior to insertion, each of the port extensions was formed into a bell-mouth. This was achieved by softening the end of the pipe with a heat gun before it was forced down onto an appropriately-sized inverted china bowl.

The port dimensions were specified by the BassBox software program used to design the enclosures. Extensive testing of the final boxes shows that the port diameters could have been made less than 100mm (with an appropriate change in lengths to match) as, at least in my use, air velocities in the ports are very low.

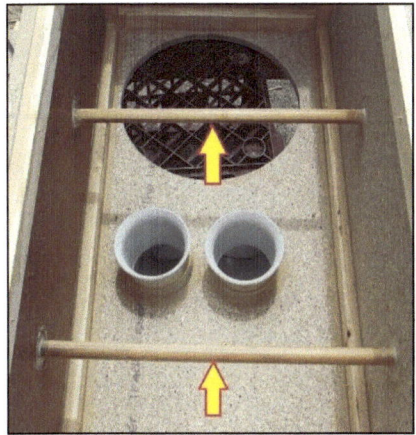

To reduce panel vibration, two internals braces are used (arrowed). These connect the largest side panels, with one brace one-third of the way along the panel and the other two-thirds along the length.

All internal walls are covered with polyester quilt wadding, glued into place with Liquid Nails. In addition, two larger pieces of wadding are rolled and then inserted into the box, one at the end furthest from the driver and the other immediately below the driver.

I chose not to use a speaker terminal; the cable simply being run out of a hole in the box that was then sealed.

Thick foam rubber self-adhesive strip was applied to the top parts of the extension pieces to form an air-tight seal with the underside of the floor around the grille.

The speaker holes were cut with a jigsaw.

Initial testing

A frequency generator was used for initial testing. I used a standalone instrument I already had, but if you don't have such an instrument, free and/or low cost downloads are available to turn a PC sound card or Apple i-device into a frequency generator.

The frequency generator should be connected to an amplifier and the amplifier used to drive the speaker under test. Be careful when driving speakers in this way: the power levels should be kept relatively low.

Testing with the subwoofer enclosure positioned in the lounge room (as opposed to under the floor at this stage) showed that there was audible bass down to 25Hz, and strong bass from about 35hz. The response was smooth up to at least 500Hz – high enough for the in-wall speakers to take over the load. This is also the time to look for air leaks – but primarily because lots of glue was used during construction, there weren't any air leaks in my boxes.

You can also use the frequency generator to 'run in' the driver, using a variety of low frequencies at different power levels.

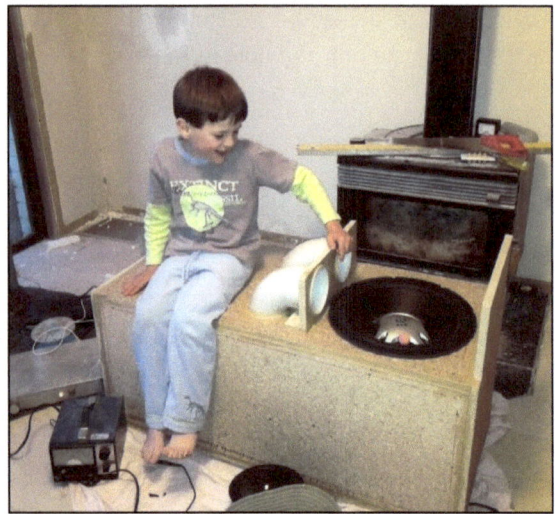

Testing one of the completed subs. My son Alexander is watching a small ball dancing on the driver, with the sub being driven by a frequency generator. Mounting the sub under the floor means much bigger enclosures can be used than would normally be permitted in a lounge room!

Foam sealing tape was used between the enclosure and the underside of the floor.

Under the floor

To provide an opening through which the drivers could fire, holes were cut in the floorboards. Because the floor was to be tiled, cement sheets had already been laid on top of the timber – so the holes were cut through both the sheets and the floorboards. Tiles were later laid around the holes.

Steel floor grilles were then placed over the holes.

To be honest, I wasn't looking forward to getting the enclosures mounted under the floor – the space is limited and the boxes heavy and unwieldy. However, with lots of help from my (then) 8-year-old son Alexander (who loves being under the floor!), things went pretty smoothly.

Cutting one of the holes in the floor. The floor was being tiled, and the hole was cut through both the floorboards and the cement sheet that had been laid.

This was the sequence of events for each box:

Step 1: We slid the enclosure along the ground on a long, narrow piece of scrap particle board, until the enclosure was located directly under the floor grille.

Step 2: Step by step, we then lifted each end of the enclosure, placing bricks under each end as we went.

Step 3: A brick was then nestled into the dirt under the middle of the enclosure, a piece of strong timber placed under the enclosure and then a surplus scissors-type car jack placed between the brick and the added timber support.

Step 4: The jack was then wound up, lifting the enclosure into place. The enclosure was raised until the rubber seal of the extension piece contacted the underside of the floor, and then adjusted up another 5mm or so to give a positive contact. (Don't wind the jack too hard: it will crack the floor!) Some additional spacers of particle board at the opposite end of the box to the speaker kept the enclosure levelled.

The jacks stay in place: with heavily greased threads, they will be useable should the enclosures ever have to be lowered for repairs or replacement. (Scissor jacks are available from wreckers and the shops at municipal tips for very little cost.)

One of the two grilles in place. Beneath lies a 15-inch subwoofer in a 200 litre ported enclosure!

The view from the crawl space underneath the house. The enclosure is lifted into place using an old car jack – the bricks under the enclosure were used to give sufficient room for the jack to be inserted. The jack remains in place.

This view shows the completed sound system. The subwoofers lie beneath grilles just able to be seen at the left and right, while the mid-bass and treble are provided by in-built wall speakers able to be seen near the ceiling.

The results

So how do the subs sound? It all depends on how silly you want to get…

With 'silliness' turned right up, the windows shake, the walls shake, the floor shakes and the lounge (on which you're sitting) shakes!

But if want to listen to music (and so you turn the sub element of the mix right down), it sounds seamlessly superb. It's just *there* – bass when there's bass in the music, and no bass when there's no bass in the music. (As opposed to systems that have resonant one-note bass that intrudes whenever the music goes below 150Hz.)

If you have room under the house (or in the well-supported ceiling), this is a brilliant way to go.

Home speakers on the ultra-cheap

> - Speakers for the shed and kids' rooms at nearly no cost
> - Improve the sound of salvaged speakers cheaply and effectively

One area of consumer electronics that hasn't fundamentally changed over the last 40 years is the design and manufacture of speakers. Whether they were originally connected to a record player, tuner, cassette deck or CD player, boxed speakers use much the same technology. So the speakers you can now pick up at garage sales, the tip or second-hand are still very useful, no matter what music source you're now using.

But nothing sounds worse than a really poor speaker – so why bother sourcing cheap or no-cost discards? Well, firstly, there are some very good speakers out there just waiting to be found, and secondly, if you have a half-reasonable starting point, it's not hard to make some major improvements while spending very little extra money.

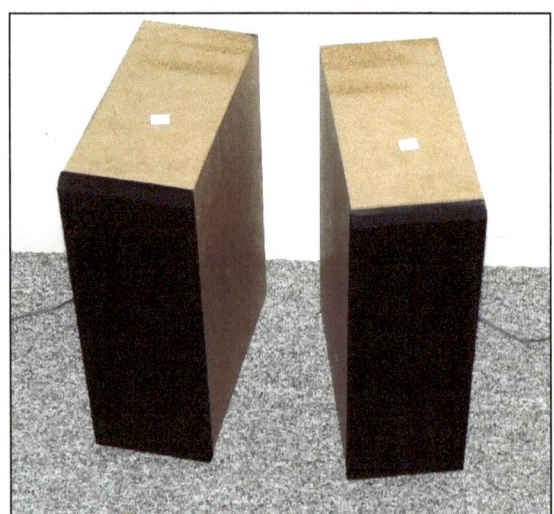

This pair of speakers was picked up at a local thrift shop for $10.

Buying

In most cases you won't have a chance to listen to a speaker that you're collecting, so how do you make any judgements as to how good it will sound? Here are some buying points:

- Pick them up and feel their weight. In nearly every case, heavier = better.

- Detach the grille and inspect the cones. The roll suspensions should be intact and the bass driver should be able to have its cone moved back and forth manually, without any voice coil binding occurring. If you cannot detach the grille, be wary.

- Either a ported or non-ported design is fine, but in the case of ported speakers, the port diameter should be large enough that whistling of chuffing won't occur. In other words, a tiny port diameter with a large diameter woofer isn't a good sign. Very large diameter (but short) ports are also unlikely to be indicative of a good design, as they'll be tuned to too high a box resonant frequency.

- Check the brand and labelled specs (eg impedance and power handling). Often the specs aren't very trustworthy, but the better the brand, the more the figures can be believed.

- Make sure that you will be able to later open up the enclosure, either by unscrewing the drivers or by detaching the back.

- Assess the condition of the boxes.

Improvements

Once home, the first step is to listen to your newly acquired purchases. Hmm, sound pretty bad? But what specifically is bad? Is the treble over-bright? Is the treble dull? Is the bass lacking, or perhaps all one-note? Try the speakers on voice as well as different sorts of music – listening to the human voice is surprisingly good way of assessing the mid-range response. PC frequency generator software is available freely on the web and it's well worth downloading some then driving your amplifier and newly-acquired speakers across a range of input frequencies.

And look, if the speakers sound absolutely awful, chalk the episode down to experience and go find some more! But if they have potential, there's plenty you can do to help realise it.

Inspection showed a decent small woofer and cone-type tweeter with a single capacitor crossover. But what was most interesting was the odd port design and the gap around the tweeter!

- **Over-Bright Treble**

Place a resistor in the feed to the tweeter. For example, 8.2 ohm ½ watt. Try some different value resistors and you'll soon get a feel for the changes that can be made. (Also see page 53 for more on matching tweeter levels.)

- **Poor Treble**

Replace the tweeter. Unless you fluke a direct drop-in replacement, this is often most easily achieved by cutting another hole in the baffle and installing the tweeter in a new spot. The old tweeter can then just be electrically bypassed.

If the grille cloth is dense and the treble improves with the grilles off, replace the cloth with a design that is more open-weave.

(Just go to a dressmaking supplies shop and buy black fabric that is easy to see through when stretched.)

- **Coloured Midrange**

In non-ported designs, place a loose fold of quilt wadding (or fibreglass insulation) inside the box. Aim to fill about 75 per cent of the volume. In ported designs, staple a thin layer of quilt wadding to the internal panels, making sure you don't block the port.

(As with grille cloth, quilt wadding is available very cheaply at dressmaking supply shops.)

- **Poor Bass**

In non-ported designs, three-quarters fill the box with quilt wadding, as described above. Also, when the speakers are working hard, use a moistened finger to check for air leaks, especially around the terminal block and the edges of the woofer.

In ported designs, try changing the length of the port. Place a rolled-up cylinder of thin cardboard in the port and move it back and forth within the port to effectively lengthen the port by different amounts. Use the frequency generator software and your PC and make lots of listening tests.

The aim is to reduce any bass resonant peaks – say, over the range from 30 – 150Hz. When you have found the right length, glue the cardboard in place. It's easy to use a spray can to paint the insides of the new port black – no-one would ever know!

- **Speaker Overloads**

If the speaker is easily driven into bass distortion, fit a 200uF non-polarised capacitor in its feed. This will drop the amount of bass being fed to the speaker and is an ideal approach if you have other speakers in the system (eg a subwoofer) to provide the required bottom end. It also works well if you're using the newly-acquired speakers as extension speakers, but still want the main speakers to be powered at high levels.

Given the relatively large box volume, I decided to alter the design into a sealed enclosure. The drivers were removed and a piece of scrap chipboard placed over the tweeter openings and port. It was screwed and glued into place. The tweeter was then re-installed from the front, the gap around its rear magnet assembly closed off with sealant.

- **Cabinets**

Unless you have got yourself a really high quality design, it's usually not worthwhile spending hours improving the finish of dilapidated boxes. However, one cheap, quick and easy approach is to give the box a quick rub back (or if it's a plastic finish, wipe over) and then spray-paint the box flat black. It won't come up with that famed 'piano' finish, but the poor surface will also not stand out nearly as badly!

- **Placement**

This one of course applies to all speakers, not just cheaply obtained rejects. The sound that the speaker makes can be dramatically altered by its room placement. If they lack in bass response, put them in the corners of the room. If the bass is strong and muddy, bring them out of the corners or even put them on stands. If the treble is muted, raise the speakers so that the tweeters are at ear level when you're seated. Always try moving speakers around – if you haven't done this before, you'll be amazed at how much you can vary their sound.

Some black spray paint concealed the blanking plate and the changed tweeter mounting.

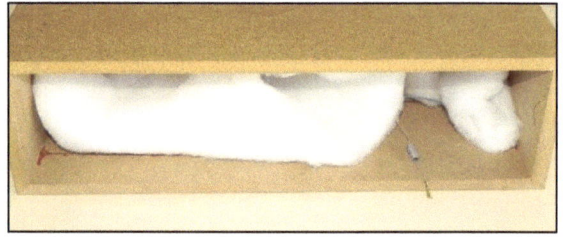

More internal fill was added (old quilt wadding was used) and the original fill was replaced in the enclosure.

The grille cloth was reinstalled and the baffle glued back into place. Comparing the modified and unmodified speakers showed a clearly better, less hollow and more natural sound.

A complete speaker makeover

- High quality bookshelf speakers for less than the price of a single woofer!
- Completely revising the internals of existing good condition enclosures

On the previous pages I improved an existing speaker system by changing the design from a ported enclosure to a sealed one, while retaining the existing drivers and the crossover capacitor.

However, this time I am making much more major changes.

The starting point was a pair of old Realistic speakers. The 6-inch, two-way design uses a cone-type tweeter.

Starting points

The starting pointy was a pair of old Realistic (Tandy) speakers. I've always had a soft spot for these speakers – walnut-veneered particle board enclosures, usually of the sealed design. The 6-inch, two-way design uses a cone-type tweeter. I got the pair for nearly nothing at the shop at the local municipal tip – they were in immaculate condition.

However, when I got them home and listened to them, I was a bit disappointed at the lack of bass. From sealed enclosures about 33 x 21 x 12cm (approximately 8 litres) you don't expect a lot of bass, but with the standard speakers there was nothing at all below 100Hz.

I removed a woofer and, using a frequency generator (an iPhone running an appropriate app) and an amplifier, I measured its resonant frequency.

(This can be done by ear alone – just scan the frequency up and down in the range from 30 – 150Hz. At one particular frequency – the resonant frequency – the driver will suddenly get louder.)

Doing so showed that the bare driver had a resonant frequency of about 115Hz.

That's very high for a 6-inch driver that is supposed to be a woofer - no wonder there wasn't much bass response!

Woofer change

Because of the high resonant frequency of the standard driver, I decided to upgrade the woofer. I had a pair of ex-JVC salvaged drivers of the same diameter and with the same mounting holes.

Tested in the same way as the original woofers, these had a free-air resonant frequency of about 70Hz – much better.

I placed the new driver in the box and listened. Bass was better – but could it be better again? Could a ported box be used instead of the sealed design? I removed the tweeter and used this opening to test different port lengths and diameters. The easiest way of doing this was to simply physically hold different ports over the tweeter opening.

Clearest changes could be heard when using the frequency generator and amplifier to test the enclosure in the range from 50 – 100Hz.

Tested in this way, best results came from a 35mm diameter port that was 90mm long.

I used a salvaged port that had a large front flare, so covering the original tweeter hole.

Tweeter change

More listening revealed that the cone tweeters were also not very good. I trialled some different salvaged tweeters and found the best results came from some soft dome tweeters I had. (During this testing I used a simple 2.2uF capacitor as the crossover.)

The dome tweeters were then mounted in new holes cut in the baffles.

But while the treble was good, it was far too bright – the tweeters being much more efficient than the woofers.

Modified enclosure on the left – new woofer, new dome tweeter and new port. Despite its appearance, the port is only 35mm in internal diameter.

Crossover

Also in my stock of second-hand, salvaged and discount purchases were some MB-Quart crossovers. These have adjustable tweeter levels of 0 dB, -2dB and -4dB. I ran the cables for the tweeter and woofer out of the back of the enclosure and then tested the sound with the MB-Quart crossovers.

However, even on the -4dB setting, the treble was still far too strong.

I then went back to the simple 2.2uF capacitor crossover, but this time trialling different resistors across the tweeter and in the feed to the tweeter. Using strings of 1-ohm resistors, it was easy to make up different values and test their effect on the sound.

Best results came from 12 ohms in parallel with the tweeter and 5 ohms in series with the tweeter.

I mounted the two resistors (they drop tweeter output) and the crossover capacitor on the back of the enclosure. These components could have gone inside the boxes, but I'd already run the wires out through holes so that I could experiment with different component values. A plastic panel covers the original terminal block.

Comparisons

I rigged up a changeover switch so that I could compare the new with the old.

First the bad news. The old speakers were much more efficient than the new ones – so with the new speakers, more amplifier power is needed for a given SPL.

However, that is the *only* bad news! The bass is better, the treble is better and the imaging is better. It's like comparing $250 speakers with $50 speakers.

Total cost, including the prices paid for the salvaged components and the speaker enclosures, was about $35!

An amplified compact speaker system

- Amplified stereo speaker for portable device
- Surprisingly good sound at low cost

This stereo, amplified speaker system has excellent sound and yet is cheap and easy to make. It's ideal working with a phone, as here. Seen on the top panel is the rotary on/off switch and a 'power on' LED.

This project came from a requirement that nothing off the shelf could quite match. We live in country Australia, only about 80 kilometres from Canberra – Australia's capital – but still in the land of kangaroos, lots of sheep farms, and a school bus that takes well over an hour to get my 11-year-old son, Alexander, to the school at the local biggest town. To while away the time, he reads books, reads (and plays games) on his tablet, watches the kangaroos and the kookaburras and the rabbits – and also, at times, wants to play music from his phone.

And Adam, the bus driver, is fine with some music.

But how to amplify the music from the phone, in a device that can also be carried in a school bag all day? Sure, there's a myriad of amplified speakers around, but is there a set that is compact, sounds reasonably good – and can satisfy the 'bus cred' that the statement "My Dad made this!" satisfies?

Not really, so I set to work.

The starting point was some marine-quality plywood, 7mm in thickness. This material formed the walls of the speaker enclosure. The front and back panels (also in 7mm marine ply) are attached to 18 x 18mm timber cleats that run around the inside of the enclosure. The outside dimensions of the enclosure are 210 x 110 x 120mm (width x height x depth).

I mitred all of the corners - and I wish I hadn't bothered going to the extra trouble as butt joints would have perfectly adequate. Marine ply (as opposed to normal plywood) was used to give as much stiffness as possible to the enclosure, while keeping it light.

The enclosure was sized so as to just fit in a stereo pair of 3 inch speakers – the enclosure has an internal volume of about 1.2 litres. The speakers were salvaged from a discarded home-style i-Phone amplified speaker system. They have large magnets and flexible roll surrounds. I could have developed a custom ported enclosure for these drivers, but measuring the Thiele-Small parameters of unknown speakers can be avoided if you're prepared to have some trade-off in ultimate low frequency response by using a sealed enclosure.

Two 3-inch speaker salvaged from a defective home iPhone sound system are used, along with a cheap but very effective pre-built amplifier module.

The amplifier module that was used is available through eBay very cheaply. (Search under 'Tripath TA2024 amplifier module'.) In this application, its benefits are its price, can work down to 7V, is efficient and fairly powerful. The amplifier is mounted on the inside of the rear panel, along with a 9V battery.

The speakers are mounted through the front panel, with the panel itself glued and nailed into position. The rear panel is removable, mounting on its inside the amplifier module and the 9V battery. Through a side panel is mounted a rotary on/off switch (rotary, so it's less likely to be bumped inside a school bag) and a blue LED used as a 'power on' indicator. (I used a LED pre-wired for 12V.)

The inside of the sealed enclosure is packed with quilt wadding – very effective at improving the sound.

Inside the enclosure is placed a generous amount of acrylic quilt wadding, serving the dual purpose of preventing acoustic reflections through the cones, and acting to increase the effective volume of the enclosure. Across the front is placed a stiff metal grille, protecting the fragile cones from fingers and odd missiles located in school bags (and school buses).

The audio signal is fed to the board via a cut-off 1/8th inch stereo adaptor cord (the cheapest way of getting a pre-wired plug), with the cord protected against pull-out by a knot.

Some wood primer and red enamel finished off the box with the required flair.

And how does it sound?

To be honest, I think it sound fabulous – far better than I'd expected. Some of that is just luck – the small drivers turning out to be good ones – but the stiff enclosure and decent amplifier module also play a big part.

When the system's frequency response is measured, the response is audible to just below 100Hz and Alexander can hear the output at 15kHz. (I'm too old to have a good high frequency hearing ability!) Admittedly, there are a few minor resonant peaks and humps along the way, but it still sounds better than all but the most high-end of small and portable amplified speakers.

So it's good enough to be also used as a picnic sound system, let alone entertain to a bus-load of kids!

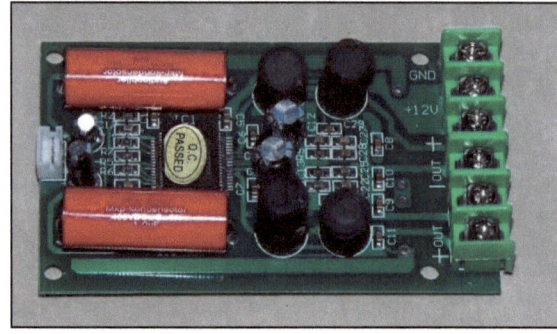

The prebuilt amplifier module is cheaply available on eBay yet provides powerful, reasonably high quality sound.

Battery life?

So, with a 9V battery, how long will the battery last? Well, that depends on how loudly the system is played.

I measured an 80mA current draw at a 'moderate' listening volume. That would make the 9V alkaline battery last something like 6 hours or so. Used 20 or 30 minutes a day, that should give reasonable battery life.

However, using a rechargeable battery (e.g. an 11.1 lithium-polymer pack and associated charger) would be a better bet if the system is to be played a lot.

Good sound from old pipes!

> - Thick-walled plastic pipe forms most of the enclosure
> - Plywood or MDF end plates
> - Can be used with sealed or ported enclosure designs
> - Quick and easy, with good sound

Building speaker enclosures is good fun - and you can also save a lot of money over buying pre-built speakers. (That's if you want quality sound anyway – poor quality speakers are available everywhere!) But while it initially seems straightforward, making good-looking, compact speaker enclosures requires excellent carpentry skills and usually a fair amount of home workshop equipment.

But here's a different approach that needs only an electric jigsaw, and is quick and easy. And, if the enclosure is designed correctly for the driver being used, the sound quality can be excellent. How good then? Far better than the vast majority of commercial 'surround sound' speakers, and much better than many speakers used in compact and portable sound systems.

So what's the trick? As you'll have guessed from the photos, you use thick-walled plastic pipe to form the majority of the enclosure. The ends are constructed from plywood or particle board.

First steps

The first step in producing such a speaker is to source a driver appropriate for use in a small enclosure.

Electronics parts suppliers sell small speakers – for example, those around 3 inches (75mm) in diameter are a good fit for the type of small enclosures being covered here. However, in order to match the driver to the enclosure, you must have either:

1. A suggested enclosure design available for that driver from the manufacturer (for example, specific recommendations for enclosure volume and / or port dimensions), or

2. The Thiele-Small specs of the driver that can be then plugged into a free on-line or commercial enclosure design software package that will calculate the required design, or

3. The ability to measure the Thiele-Small parameters of the driver yourself, with these then able to be used in a free on-line or commercial enclosure design software package.

Probably the easiest small enclosure to make is one formed from thick-walled plastic pipe, with the end-plates made from plywood or fibreboard. The sound can be surprisingly good.

Having played with DIY speakers and speaker design for a long time, I recommend the last approach. I use Woofer Tester 2 (as described on Page 10), a software/hardware package that I have found to be brilliant at rapidly measuring the Thiele-Small specs of the driver, and then designing an enclosure to suit.

A huge benefit of measuring the specs yourself is that you can source drivers from anywhere – the ones used in this story were picked up very cheaply second-hand as home theatre surround sound speakers. In the original enclosures they sounded terrible; in the new custom designed enclosures, they sound excellent! (More on this later.)

Doing it – first design

Let's take a look, step by step.

The salvaged speakers that donated their drivers were these small and poor-sounding Sony surround sound units.

The first of the two enclosure designs I am going to cover uses 70mm drivers taken from 'cube' speakers - in fact, the rear speakers of a Sony surround sound system. Each original enclosure was only about 80 x 80 x 100mm – at about 0.6 litres, small indeed! A tiny rectangular port about 25 x 5mm was located on the rear.

Four of these speakers were bought for AUD$10 second-hand – and that price was for the lot!

And how did I know these drivers would be of sufficient quality to work well in a better designed enclosure? I didn't! But at that price and with the ability to easily measure the Thiele-Small specs, it was worth taking a punt.

In fact, the measured specs were:

- Impedance (Re) = 2.9 ohms
- Resonant frequency (Fs) = 153Hz
- Total Q (Qts) = 1.36
- Compliance (Vas) = 0.54 litres
- Sensitivity = 81.7dB at 1 watt, 1 metre

One of the Sony drivers removed from its original enclosure.

These specs actually look pretty bad (high resonant frequency, low sensitivity and so on) but at 70mm diameter (and with an effective cone area of only 65mm), it's a very small driver!

Some modelling with the Woofer Tester software showed that best results came from a 1.9 litre volume tuned to 132Hz with a 25mm internal diameter port that was 27mm long.

With the internal diameter of the pipe enclosure being 126mm, a 1.9 litre volume requires a length of 152mm (ie cross-sectional area of the pipe multiplied by length). Add 20mm for the thickness of the two endplates, and add a bit for the volume for the port and speaker, and I went with an overall length of 180mm.

The port was formed from 25mm internal diameter plastic electrical conduit – it needed to be only 27mm long, and a building site offcut provided this.

Use plenty of building adhesive to glue the end plates into place – the seal must be perfect. Get water clean-up glue and then wipe off the excess with a damp cloth.

The end plates were cut from medium density fibreboard (MDF) with an electric jigsaw, and the jigsaw was used to cut out the hole for the speaker.

The port was placed in the other end panel and the hole for this was formed by a hole-saw.

Formed into a cylinder and then fed through the speaker opening was a piece of polyester quilt wadding (available from dressmaker supply shops). The rolled wadding springs open once inside the enclosure and so lines the interior pipe wall. It prevents reflections of sound off these internal walls and subsequently out of the port, or through the driver cone.

The driver can then be installed and the speaker frame and surrounding panel sprayed black.

A car sound grille was then installed and thin carpet added. For places where appearance doesn't matter (eg in a workshop), the paint, grille and carpet steps can be skipped at a considerable cost saving.

Results

So how does it sound?

For fun, I rigged up a changeover switch so I could make instant back-and-forth comparisons between one of the original Sony surround sound speakers and the new 'pipe' enclosure. After all, same driver - just different enclosures.

The difference was simply staggering.

The Sony surround sound speaker lacked any bass. Furthermore, the mid-range sound was distinctly coloured, presumably through enclosure panel vibrations and reflections. Switching to the new enclosure, the sound was immeasurably better. Singers' voices no longer had unnatural timbres and the music was much more full-bodied.

Testing the new enclosure with Woofer Tester 2 showed good response down to 130Hz; the original could get down to only 190Hz!

Doing it – second design

The second speaker system was produced in much the same way – as it happens, using drivers again taken from a Sony speaker system. These speakers were also bought second-hand for chickenfeed, but this time the speakers looked like they'd been part of a small desk-top sound system and used ported, particle board enclosures.

Measuring the Thiele-Small specs of the drivers and doing some modelling indicated that the original ported enclosure design was in fact quite wrong for these drivers! (So why did they sell them like this? Who knows... maybe the measuring and modelling software was simply not used – or not then available?)

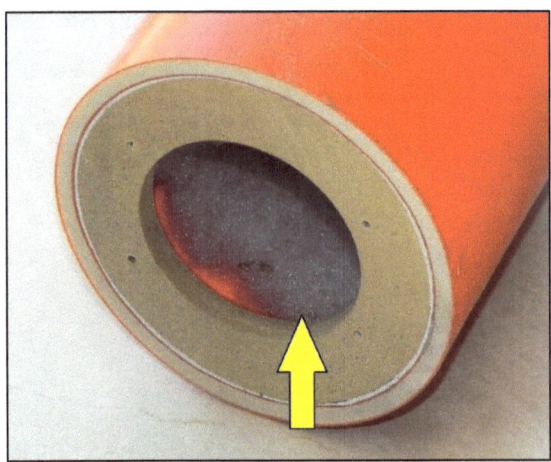

The arrow points to the quilt wadding placed inside the enclosure.

Instead of the original ported boxes, a simple sealed enclosure of 2 litres gave good modelled results. To achieve this, more pipe-type enclosures were made, but this time without a port. In addition, a larger amount of quilt wadding was placed inside the enclosure – this causes the driver to 'see' a slightly bigger enclosure, which in a sealed design is beneficial. Finally, this time I used 10mm plywood for the end plates, rather than fibreboard.

The results? Speaker efficiency is lower – that's because, all else being equal, more amplifier power is needed with sealed designs than with the ported speakers. But again, the sound 'straight out of the box' is absolutely fine. I actually built these enclosures as a weekend project with my 11-year-old son, and when the speakers were finished, he immediately wanted to run them in our home workshop.

The small speaker enclosures are almost lost from view here – they're either side at the top of the shelving. They can fill a big shed with sound.

I must say I was a bit doubtful (at 14 metres x 8 metres and with a 5 metre height, our shed is a very big one to try to fill with sound from such small enclosures) but with a bit of bass and treble boost provided by the amplifier, they were quite up to the task of providing a background radio.

Superb spherical enclosure home hi-fi speakers

- Fantastic hi-fi sound
- Use as standalones or with a subwoofer
- Spherical non-resonant enclosures give excellent presence
- Quick and easy to build

These high quality hi-fi speaker use Alpine 6.5 inch woofers, dome tweeters and full crossovers. The rear-ported enclosure is made from two clay-fibre bowl-shaped flower pots glued together. The spherical design reduces internal standing waves and diffraction effects. The stands were made from square tube.

The trouble with writing about speakers is that everyone's listening perspective is different.

The person who has listened to only MP3s played through inferior earphones – after that experience, anything is good! And, at the opposite end of the pack, the person who believes that 'tube sound' (complete with all its distortion) is without a doubt the best – what can you say?

So, where can I start?

I dunno - I don't know if it's even possible to be dogmatic in this space.

But what I do know is that, when the recording is done at the highest quality, and the amplifier is faithful to the original – well, in that case, the speakers pictured here are simply extraordinarily good.

Are they the best speakers you can buy (or make) at this price?

Impossible to know.

But what I can say is that their transparency, their ability to "be with you in the room" is quite phenomenal. Add to that their efficiency (in a normal listening room, 50 clean watts per channel is absolutely more than adequate); their frequency response range and lack of diffraction peaks and troughs; their ability to cope with bass and treble boost as you desire; and their ease of building for yourself - well, these some of the best speakers you can get for the money in the whole world…

So what, exactly do we have here?

Firstly, we have drivers that many audiophiles turn up their noses at – car sound speakers! Then – and get this – we have some flower pots being used as enclosures!

Drivers

Let's look at the topic of car sound speakers.

Many car speakers are very poor. In a way, it goes with the territory – if you're making speakers that may be randomly mounted in car rear decks, or doors, or dashboards – well, you produce speakers with high Q values, so that they give plenty of one-note bass when mounted in enclosures that range from small to huge.

But of course, that means for real hi-fi use, they're useless.

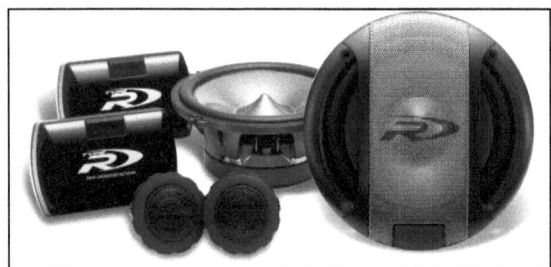

These Alpine component speakers have superb performance for the price.

But not all car sound speakers are like this. The Alpine R-series 'splits' (i.e. woofers, tweeters and specific crossovers) are excellent – in any context. The speakers shown here use the Alpine SPR-17S 6.5-inch Type R Component Splits - and I can't speak highly enough of these drivers.

The woofers have tested specs of:

- Actual cone diameter: 5 inches
- Resonant frequency (Fs): 75 Hz
- Total Q (Qts): 0.77
- Compliance (Vas): 5.9 litres
- Sensitivity: 86dB at 1 W/1m

While these specs may not seem up to the same standard of some 'premium hi fi' brand name speakers, can I say that that in the real world, these Alpines are very, very good.

The bass drivers have specs that lend themselves to compact ported enclosures, and the crossovers have variable tweeter attenuation so you can tweak the high response as you wish. The drivers also have plenty of power handling and, since in ported enclosures their efficiency is fairly high, lots of SPL is available if wished. At the opposite end of the spectrum, they also still sound good with literally only a few watts powering them.

The Alpine SPR-17S set comprises nominal 6.5 inch woofers with cast alloy frames, 1-inch dome tweeters, and full crossovers (not just capacitors!)

At the time of writing, they're still available - but are being replaced with a newer models... and so are available relatively cheaply at around AUD$215. Given that includes two good woofers, two excellent tweeters and two very good crossovers, can I say that's excellent value?

The enclosures

The enclosures are spherical fibre-clay designs, made from commercially available, bowl-shaped flower pots. So how are the enclosures constructed?

In short (more detail on the process in a moment) the flattened surfaces on which each bowl normally sits is ground smooth using an angle grinder. In one of these flats is cut the hole for the driver; in the other is cut the hole for the speaker port. The driver and port are installed in their respective bowls, then a layer of polyester wadding is placed inside the bowls. The bowls are then glued together using industrial adhesive.

The result is an enclosure that is very stiff (the curved walls flex little), and is shaped so that standing waves (internal reflections that normally occur off the flat walls of a speaker enclosure) are not present.

The enclosure can also be very quick to build (under an hour, easily) and it does not require woodworking tools or machinery.

The enclosures have an internal volume of 16 litres and are tuned to 49Hz with a port 50mm in diameter and 118mm long.

The tweeter is mounted below the main enclosure, and the crossover is mounted on the rear.

Let's take a look at the build, step by step.

The first step is to buy a pair of the bowls. At the time of writing, these bowls were available in 28cm, 34cm, 40cm and 51cm diameters, and in black or white finishes. The one used here is 34cm in diameter with a black finish. It gives a 16 litre internal volume.

A normal metal grinding disc mounted in an angle grinder does a good job of removing the 'feet'. However, it's a dusty job and so it's best done outside with the operator wearing googles, hearing protection and a dust mask.

Here is the as-bought bowl turned upside-down. While it isn't super clear in this pic, the bowl rests on 'feet' which are formed into the bowl. To gain a flat surface, these feet need to be ground off.

Here's the view with the base smoothed. To get it as flat as possible, follow-up the grinder with the use of a belt sander, or moderately coarse sandpaper and a sanding block.

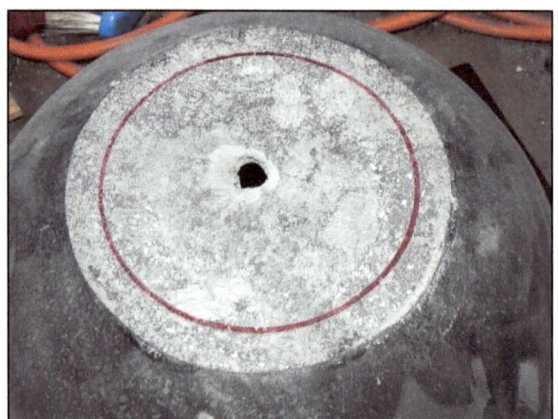

The next step is to mark the cut-out required for the driver. Be careful in sizing of this hole – too small and there will be the need for lots of further grinding; too large and you will not get the driver to seal in the hole. Note the pot drainage hole.

Here is the hole immediately after the jigsawing had been done. It's difficult in the clay-fibre mix to follow the line accurately (there are small pebbles in the material too), so it's better to err on the side of undersize rather than oversize. The hole can then be ground back to the line by carefully using the angle grinder.

The hole for the driver can be cut out with an electric jigsaw – this pic shows the piece that's been removed. Use a coarse wood-cutting blade – and expect to get only a few holes out of the blade before it is blunt. Don't go like a bull in a china shop – you can crack the bowl if you push too hard.

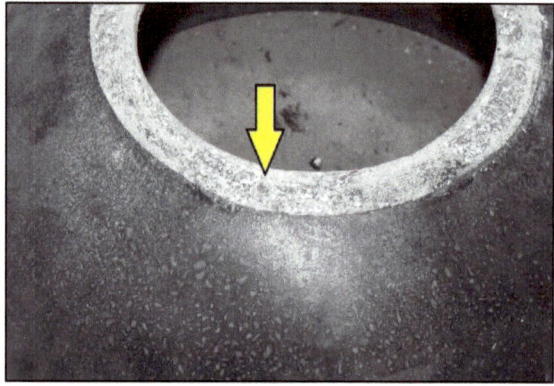

It's hard to see, but the arrow points to one of the speaker mounting holes that has been drilled. If you have a masonry drill bit, you could use that (but not with a hammer drill!) but these holes were drilled with a normal high speed steel bit. Drill the holes sufficiently large that machine screws can be inserted – nyloc nuts and washers will go on the inside of the bowl.

Use silicone sealant under the edge of the speaker frame. This is important because despite smoothing with the grinder and sandpaper, the mounting surface is likely to not be perfectly flat. You don't want air leaks around the edge of the driver.

The driver installed in the hole. Note the use of washers and nyloc nuts on the speaker mounting screws. Tighten these when the sealant is yet to set – but be careful not to over-tighten or you could crack the bowl.

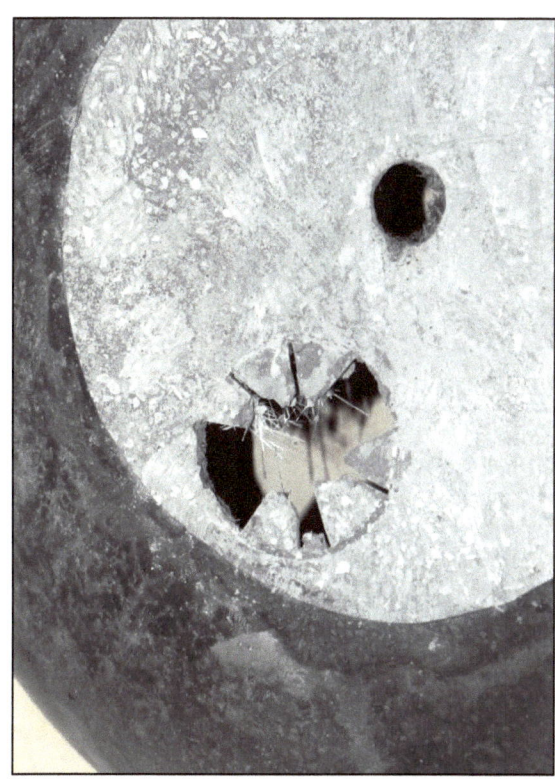

With the driver installed, it's time to tackle the other bowl. The other bowl, it's 'feet' already ground off, needs a hole made for the port tube. I found the easiest way to do this was to use the jigsaw to cut a series of radial slots like this, with the pieces then broken off by the careful use of pliers. The hole was then filed to final shape.

The hole doesn't have to be perfect – the industrial glue used to secure the port in place will fill any small gaps.

(Don't use an expensive holesaw to make this opening – the teeth will soon be blunt and the holesaw ruined.)

The port tube needs to be the correct diameter and length to match the enclosure design. PVC pipe is cheap and easy. This port is 50mm in diameter and 118mm long.

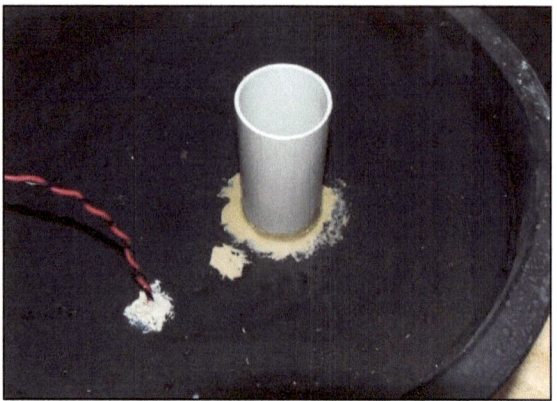

Now is also the time to feed the cable through a hole drilled in the bowl, and seal it with further glue. Either solder the wires to the woofer's terminals or equip the cable with push-on terminals to make this connection.

The port glued into place. For this task, and also gluing the bowl halves together, I used water clean-up Liquid Nails. Note that this glue has also been used to seal the 'water drainage hole' provided in the original pot.

This photo actually shows two steps. The first is that polyester quilt wadding has been placed around the inside of the lower bowl. Ensure that there is enough projecting upwards from the lower bowl so that when the two bowls are glued together, the wadding covers the inside surfaces of the upper bowl as well. Also ensure the rear of the driver and the port are not blocked by the wadding.

The second step is to glue the two halves together. Use a generous amount of glue around the rim of the lower bowl. The upper bowl is then carefully lowered over the lower bowl, ensuring the wadding goes inside the upper bowl as it is lowered. (This job is best done by two people.)

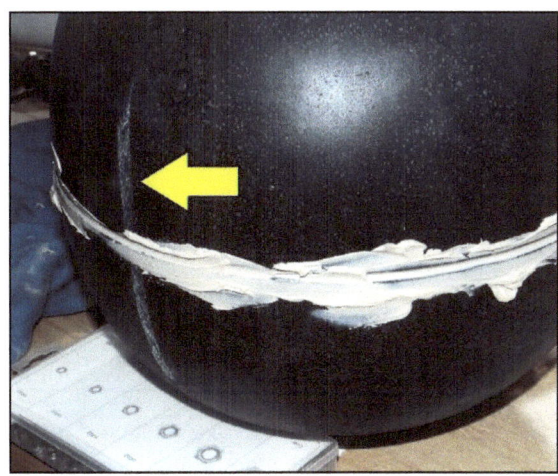

The bowls are pushed together and then a wet finger used to smear the glue around the join, ensuring that the gap is completely filled. A damp cloth is then used to carefully remove the surplus glue. (The glue must be of the water clean-up type!) Note the chalk witness line (arrowed) that shows the correct alignment of the bowls as they are joined. This line is made earlier when the best rotational fit is found – because their lips aren't dead-flat, the bowls fit together better in some orientations than others.

The tweeter is mounted below the main enclosure. I could have run the tweeter cable within the square tube, but in normal use the cable is not as visible as here. The physical position of the tweeter makes a substantial difference – it must be as far forward as the main driver.

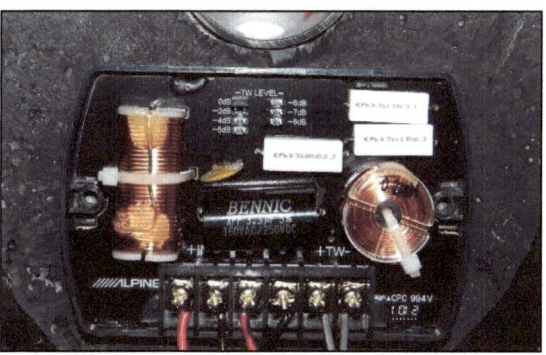

The next step is to mount the crossovers on the rear 'flat'. This was glued into positon just below the port.

The stand shown here was made from square tube. You can make any type of stand you wish – but the speaker must be off the ground to work at its best.

Testing speakers

- Frequency response testing with your ears
- Frequency response testing with a phone app
- Comparative listening tests

If you are building or modifying speakers, you will need to test your results. That's especially the case if your development is a step-by-step process.

(Note that if you are using new drivers, you must first 'run them in' for a few hours before they'll develop their final sound.)

Frequency response testing (using your ears as the sensor) can be done using a frequency generator like this one, or a phone app or PC software.

Frequency response testing using your ears

Using a frequency generator (or an appropriate app on a phone or PC) and an amplifier, do a frequency sweep. That is, vary the sound from a very low frequency like 20Hz right up to 20kHz.

However, be sure to not over-drive the speaker, especially at low frequencies!

When listening to a frequency sweep, is the response smooth, or are there obvious peaks and troughs? Be particularly watchful (earful?) at low frequencies where inappropriate enclosure design can result in loud, one-note bass.

What are the lowest and highest frequencies you can hear from the speakers? However, note that as you get older, your high frequency hearing declines – so irrespective of how good your speakers are, you may not be able to hear them at high frequencies (eg above 10kHz).

In ported designs, is there obvious port noise ("chuffing") at low frequencies? By placing your hand on the enclosure walls, at low frequencies is there obvious enclosure wall flex?

A simple frequency response check is extremely valuable.

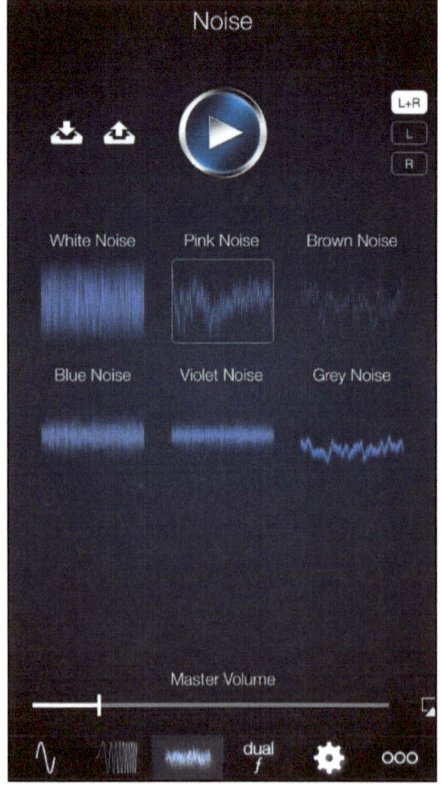

Different 'noise' (eg white and pink noise) can be generated using a cheap phone app. This one, Function Generator Pro, can also do frequency sweeps – so replacing the instrument shown at left.

White noise testing

White noise is noise that has equal energy on all audible frequencies. As it is 'treble rich' to the ear, it is particularly useful when setting-up tweeters - minor changes in crossover frequencies and tweeter levels can be easily picked.

Pink noise has equal energy in each *octave*, and so has the same apparent loudness to the ear at all frequencies. Pink noise testing gives you a better feel for how the speakers will sound in real life.

Both white and pink noise apps are available for phones or as downloadable PC software.

Pink noise is particularly useful when making comparisons between two different speakers – more on this in a moment.

Frequency response testing using an app

There is a multitude of hardware and software combinations that will allow you to measure the frequency response (and other aspects) of your speakers. How accurate you want to be depends on how much you want to spend!

I choose to use the Studio Six Digital FFT app that works with an Apple i-device. This software claims to switch off some of the normal changes that Apple program in the frequency response of the microphone, and so give good results at low cost (if you have an Apple i-device, anyway!)

In use it will clearly pick changes being made to tweeter levels, and dips in the response curve that you can also confirm with your ears by doing a frequency sweep. However, without listening in an anechoic chamber, you'll also get lots of room reflections – so it all starts getting a bit hard to sort the wheat from the chaff.

Also note that moving the microphone only a very short distance relative to the axis of the speaker can change the measured response curve quite significantly – that is, it's easy to get a smoother curve by positioning the microphone just-so!

Given that we all listen in rooms, and that we don't stay rigidly in a specific sweet spot (that can change by moving only centimetres) it's likely that low cost speaker measurement is more fun than helpful – especially if you are adept at picking problems with your ears alone.

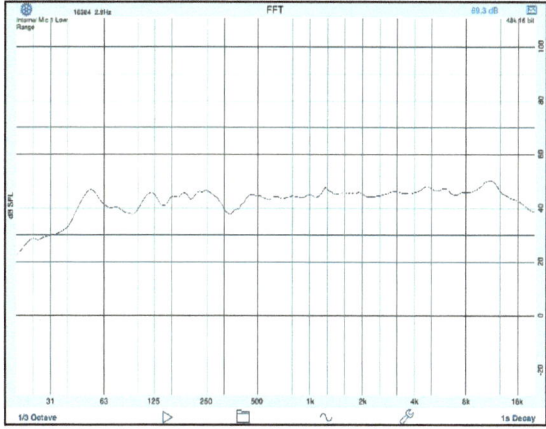

Here is a 1/3rd octave frequency response curve of one of the spherical speakers shown on page 37, as measured by the Studio Six Digital FFT app working on an iPad Air. It shows a response of plus/minus 3dB from about 45 to 17000Hz, except for two dips – one at about 350Hz and the other at about 100Hz. These dips are quite likely to be caused by room reflections.

Listening tests

Before making listening judgements about your speakers, you need to know what sound is actually good, bad and indifferent.

Perhaps the best music reproduction that you're able to easily access is that provided by quality headphones. Many shops selling headphones are happy to let you listen to them - something they tends not to do as much with expensive speaker systems!

Listen to your favourite tracks on quality headphones and develop a feel for what is good. It's more than just treble and bass and midrange – it's the ability to hear the location of instruments across a sound stage, and to immediately pick good recordings from bad.

I find the last difference (differentiating good recordings from bad) the quickest way of knowing that I am listening to good speakers. If you're listening to good speakers, you'll be able to make judgements of how well the vocals have been placed in the mix, and how close or far away the different elements of the music are. For example, the spherical speakers, covered in the previous section, have extraordinary 'presence' – the singers and instruments sound right 'in your face', not five metres away.

Good quality headphones can be used to 'tune' your ear to recognise the nuances of good quality sound. You can often listen to even quite expensive headphones in shops.

Listen also to audio material on your speakers that perhaps you would *not* normally listen to. For example, the spoken voice is very good at allowing you to spot mid-range colouring, or sibilance (peaky treble).

If you are doing a lot of step-by-step development, listen *in the room that will finally house the speakers*. The size of the room, the furnishings and floor coverings - all have a quite major effect on the sound quality. In fact, in our house, I can hear the difference in the sound of the speakers when our lounge room has been tidied!

All speakers vary in their sound with their proximity to walls, the floor or the ceiling. However, some designs are extremely susceptible to major changes in sound with these differing locations.

For example, the in-wall speakers described on page 13 changed dramatically in sound with location. In fact, they were at their worst when placed halfway between the floor and the ceiling – the intended location! The spherical speakers described on page 37 must be on their stands to sound good, and furthermore, 'prefer' to be located away from walls as well.

Ensure that you ask other people to also make listening judgements. While some of us might have 'golden ears' that can remember how a system sounded before a change in (say) the crossover frequency, most of us tend to be swayed by what change we've just made!

Added stuffing to the enclosure? Now doesn't *that* sound better! Having another, more impartial, person make a judgement is very useful.

Make comparisons between different speakers. The best way to do this is to drive the two different speakers from the one source. Then, using a single pole double throw (SPDT) switch, flick back and forth between them.

However, if the speakers vary in efficiencies, be careful that the louder of the two speakers isn't automatically regarded as better. In that case, you need to adjust the volume so that the two speakers being compared are equally as loud. Wiring a high power L-pad into the circuit will allow you to reduce the volume of the louder speaker so that it is match for the other.

When doing this 'switching' comparison, quite subtle changes can be detected. For example, I recently compared two small bookshelf speakers that I'd acquired.

I first analysed the standard bookshelf speaker design using Woofer Tester 2. That is, I removed the woofer and measured its specs, then plugged into the software simulator the actual enclosure specs – ie both volume and port dimensions. Doing this showed that the manufacturer had done a pretty good job – and in fact I thought the standard speakers were not too bad.

However, opening-up the boxes showed a lack of any wadding, so I added some to the internal walls. I also shortened the port a bit, something that the simulation software had indicated would lift the bass response just a little.

These are small changes – but in doing a blind 'switching' comparison, both my wife and I repeatedly picked the 'improved' speaker as sounding better.

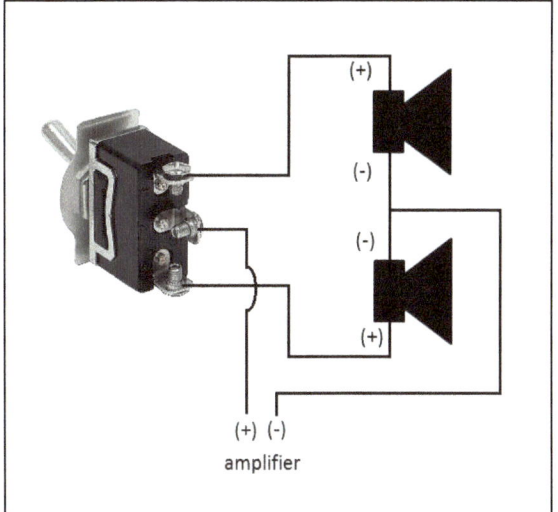

Making a comparison between two different speakers is easy if a single pole, double throw switch is used. As shown here, the two speakers are wired to the one channel of the amp. Flicking the switch changes the speaker that is selected. Single drivers are shown here but the two selected speakers can be complete speaker systems.

I also suggest that you do all your testing using the signal source and amplifier that you will be running in the final set-up. For example, the ability (or not!) of the amp to provide high current gulps can make a major difference to how punchy the bass sounds.

Note that, having said that, I often use a cheap amp module for preliminary testing. How cheap then? Try (at the time of writing), under $17!

The amp is available on eBay (search under "TPA3116 High-Power 2.1 HIFI Digital Subwoofer Amplifier Verst Board 50W+50W+100W") and needs a 24V DC supply. (I provide this from my bench power supply.)

Because it's a 2.1 amp (ie a stereo pair plus subwoofer), it's good when you are testing a sub with two satellite speakers, or a stereo pair. (Just connect what you want to listen to.)

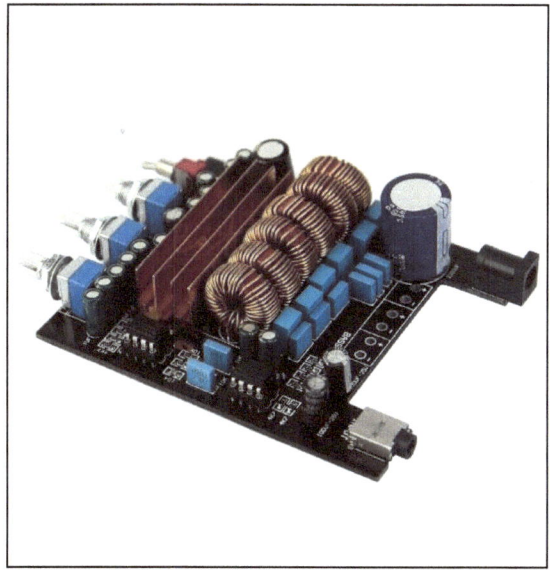

I often do preliminary testing with this 2.1-amp module. It's rated at 50W x 2 plus 100W x 1 and is a good 'test' amp to use as you develop speakers. Cost is low – but it does need a 24V power supply.

A final speaker test is to listen at low volume for a few hours while you do something else. Some speakers can be quite wearing, while others you can listen to all day. (I assume that this difference relates to subtler aspects of distortion.)

Woofer Tester box testing

The Woofer Tester 2 hardware / software combination described on page 10 can be used to test the behaviour of a completed loudspeaker enclosure – a "box test".

This is useful when, for example, you want to see if the actual impedance curve matches the modelled impedance curve. In turn, this shows aspects such as the resonant frequency of the finished speaker system.

Gaining quality loudspeaker drivers at low cost

- Stripping damaged or unwanted hi-fi speakers of their drivers
- Salvaging drivers from cheaper mini-system and home theatre speakers
- Car sound system speakers, including original equipment

The major cost associated with building your own speakers are the drivers that are used. As described earlier in this book, by the time you buy new woofers, midranges, tweeters and crossovers, you're already likely to be paying more than you would for similar quality, fully-built speakers purchased from a retail shop!

So how can you get drivers – woofers, midranges and tweeters – cheaply?

It's actually pretty easy – just buy second-hand speakers, or get the drivers from speakers that people throw away.

Let's take a look at doing so.

Superficial damage like this can dramatically drop the asking price for high quality hi-fi speakers. If you're after just the drivers, the damage doesn't matter.

I have often found JVC drivers (like this one) to be of good quality. This 5-inch woofer, salvaged from a compact hi-fi system, has a measured Fs of 68Hz, Qts of 0.48 and Vas of 6.4 litres. In an 8 litre ported enclosure tuned to 56Hz, it has a modelled -3dB point of about 50Hz – and that's what it matched in reality! Cost for the driver was $5.

Complete hi-fi speaker systems

As I write this, on eBay someone is selling a pair of Dali Blue 5005 speakers for $150. These speakers each use two 6.5 inch bass / midrange drivers and dome tweeters in a ported enclosure. To put that another way, for $150 you get six high quality drivers and two crossovers – that's under $19 each!

And why are these quality speakers so cheap? The cabinets have some cosmetic damage – that's all. Of course, if you don't intend to use the original cabinets, then that is no problem at all.

Earlier in this book, I used the 8-inch woofers, dome tweeters and crossovers salvaged from a pair of Wharfedale speakers. They were used to build new in-wall ported speakers that sound excellent.

So, buying complete, 'brand name' speakers that are unwanted or have some cosmetic damage to their cabinets can give you high quality, matched drivers very cheaply.

Complete lower quality speaker systems

You can also move away from high quality speakers and buy – even more cheaply – second-hand speakers used with small hi-fi systems and home theatre systems.

My nearest large city has a shop that is located next to the municipal rubbish tip. It sells all small cabinet speakers at $10 a pair – irrespective of their brand. Larger enclosures are $15, and floor-standing enclosures are typically $25 a pair.

Even cheap speakers often have salvageable parts like these flared ports.

Most cone tweeters able to be salvaged from cheap speakers are poor in performance. These, however, are an exception, with quite satisfactory treble for surround sound, home theatre or similar applications.

At this shop these speakers can be inspected but not tested. There are no guarantees as to the quality of the speakers – or even if they work at all. However, if selected carefully, about half of the speakers I have bought in this way have had bass drivers of sufficient quality that they can be used in good home-built systems.

Note that I said 'bass drivers'; my experience is that the cone-type tweeters used in these cheap systems are quite poor. Note also that small dome tweeters used in cheap systems typically cannot be salvaged – they're glued in place and cannot be removed without damaging the tweeter.

All of these low cost speakers use very simple crossovers – typically, with just a capacitor and a resistor being used to feed the tweeter.

Whether dealing with either genuine hi-fi speakers or cheaper low quality designs, I suggest that you prefer ported over non-ported enclosures. Furthermore, try to select enclosures where the manufacturer has used properly flared, circular ports.

This has two advantages – the ports can be salvaged for re-use, and the more careful design carried out by the maker usually indicates that the speaker is a better design overall.

Many hi-fi speakers have crossovers that can also be salvaged. Some, like the one shown here, are integrated with the rear terminal block.

The woofer from the Alpine SPR-17S car 'splits' system is a quality unit. Not all aftermarket car sound speakers are this good, though!

Car speaker systems

There are many car speakers on the market – both brand new and second-hand. Let's start at the most expensive end first.

Aftermarket car 'split' systems (comprising woofers, tweeters and separate crossovers) can provide excellent, matched drivers. The spherical speaker system shown on page 37 takes this approach and uses Alpine drivers. However, Thiele-Small specs are never available for small car sound drivers – so you need to measure these parameters for yourself. And of course, to measure them, you usually need to have bought the speakers!

If you do this Thiele-Small testing, you will find that many car sound speakers have relatively high resonant frequencies and high Q values. These characteristics give the drivers plenty of (one-note) bass response when mounted in a wide variety of sealed enclosures – for example, in a car door or boot (trunk). However, such driver characteristics don't lend themselves to smaller ported enclosures being used for quality sound use. (And the sealed enclosures that are required are usually too large.)

However, higher quality car splits (such as the aforesaid Alpine drivers) do have good measured specifications.

So where does that leave you? With aftermarket splits, the bass driver may not be suitable for mounting in relatively small enclosures, especially ported ones. However, that still leaves the tweeters and separate crossovers – both of which are usually excellent. Most splits use dome tweeters and 'full' (inductor, capacitor and resistor) crossovers, with the tweeter level often user-adjustable in output loudness.

Car subwoofers normally **do** have the Thiele-Small specs available, so you can make a more informed choice – whether buying new or second-hand.

Note that typical low-cost, wide-range car sound speakers are usually not worth using.

The other source of drivers from car sound are original equipment speakers. These often appear for sale as used items – the car owner has upgraded the original speakers and is selling off the old ones cheaply.

Premium and luxury cars have for years been using high quality drivers, often mounted in ported enclosures. For example, Lexus invariably use ported speaker enclosures in the doors of their cars. These speakers have excellent specs and so make a good buy for DIY speaker building.

This 5-inch second-hand Lexus speaker has good measured specs: Fs of 68Hz, Qts of 0.68 and Vas of 10.9 litres. In a 23 litre ported enclosure tuned to 47Hz, it has a modelled -3dB point of about 45Hz.

However, that doesn't mean that *all* Lexus speakers are good for home (or car sound) – it depends where in the car they were used. For example, the speakers originally mounted in the rear deck (and so using the whole volume of the boot [trunk] as their enclosure) have the high Q characteristics that makes them unsuitable for small domestic enclosures.

An 8-inch woofer from the rear deck of a premium car. It has a measured Fs of 43Hz, Qts of 0.52 and Vas of 35.6 litres. In a large 50 litre ported enclosure tuned to 38Hz, it has a modelled -3dB point of about 35Hz.

But again it's horses for courses. One car maker uses 8-inch woofers in the rear deck of their premium sedan. These speakers are no good in small enclosures, but a relatively large (50 litre) ported enclosure works well with them.

So what to do when sourcing original equipment car sound drivers?

Firstly, to be able to design appropriate enclosures for these speakers, you need to be able to access a testing system like Woofer Tester - or you're just blindly stumbling around in the dark.

Quality crossovers are available relatively cheaply if bought as part of a car 'split' system.

Secondly, if the drivers look good and/or have come from a car renowned as having a good quality sound system, *and if the speakers are cheap enough*, buy them. It's not hard to pick up quality drivers sourced in this way for $30 a pair – worth taking a punt.

A real world example

Recently listed on eBay, and available near where I live, was a job-lot of 17 second-hand ceiling-mount speakers. The design was an 8-inch, two-way with concentric dome tweeter, and the grille surrounds and grilles were included. Price? Just $110 for the lot!

The speakers were equipped with line-level transformers. However, the very same driver was also available without these transformers - where it was rated as an 8 ohm, 45 watts RMS driver.

I bought them and then when I got them home, I bypassed the transformer and tested one of the speakers with Woofer Tester.

Unfortunately, the results were not very good, with a resonant frequency of 53Hz (fine), but a Qts of 0.7 and a Vas of 38 litres. (To an extent, with a speaker designed to be mounted in a ceiling, this is what you'd expect.) Modelling different enclosures with Woofer Tester indicated that a sealed enclosure would need to be a rather large 50+ litres, and ported enclosure designs were not very effective because of the driver's high Qts. That's what measurement and modelling showed, but what would a listening test indicate?

I grabbed an old Akai speaker box of about 55 litres, glued and screwed a new MDF front panel to it, and installed one of the new speakers. (The resulting large, sealed enclosure best matched what the modelling had shown would be most likely to be effective.)

However, the speaker sounded dreadful! It had very poor bass and an overly-dominant mid-range. But the treble was fine.

Hmm, so what to do with 17 of the speakers? Removing the label on the back of the magnet revealed a long screw that held the tweeter in place.

Removing the tweeters (each complete with its own crossover capacitor) was therefore quick and easy. I also salvaged the grille surrounds and aluminium grilles, and chose to also keep the large magnets (that I'll use in a completely different application - to hold steel bits together when welding in my home workshop).

I bought 17 of these second-hand 8-inch 2-way ceiling speakers for just $110!

So were the dome tweeters any good?

A few hours of testing revealed that yes, they were indeed fine. I ended up using a pair, crossover capacitors bypassed, as the tweeters in a pair of speakers that use small, clay-fibre, spherical enclosures (see page 60).

(The tweeter crossover capacitors were most easily bypassed by soldering a wire in parallel with them. That way, if required, the crossover can be easily re-established in the future by just unsoldering the wire. Be careful when soldering to dome tweeters – the solder terminals can easily come right out, ruining the tweeter.)

However, despite their good appearance, their bass proved to be terrible and the midrange overly dominant.

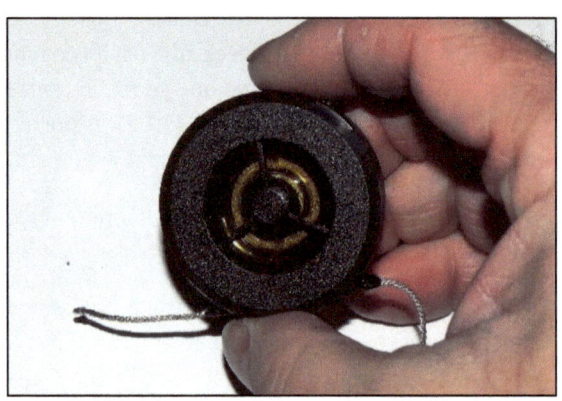

But it was easy to remove the dome tweeters, and these were quite good. Hmmm – 17 good tweeters for $6.50 each!

Adjusting tweeter levels

> - Cheap ways of turning down the tweeter level to match the rest of the system.

Invariably when setting up a speaker system, the tweeter is louder than you want it to be. That's typically the case because the tweeter is more sensitive than the woofer and/or midrange. So how can you cheaply turn it down, so that the mix is not dominated by treble?

Let's assume that you have set the crossover frequency appropriately, either by the use of a 'full' (inductor and capacitor) crossover, or by the use of a simple capacitor.

Overall approach - changing resistances

The usual approach to dropping tweeter level is to use resistors.

Simplest is to just put a series resistor in line with one arm of the tweeter feed – R1 in the diagram below. For example, you could insert an 8 ohm, 5 watt resistor. This decreases tweeter output - but unfortunately it also changes the impedance that the crossover sees – so changing the crossover frequency.

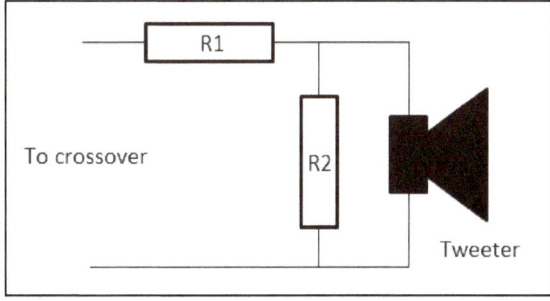

In order that impedance remains constant, two resistors (R1 and R2) are used, rather than just one. One resistor is in parallel with the tweeter, and the other is in series with it.

To keep constant the impedance that the crossover sees, two resistors are used. One is in parallel with the tweeter, and the other is in series with the feed from the crossover.

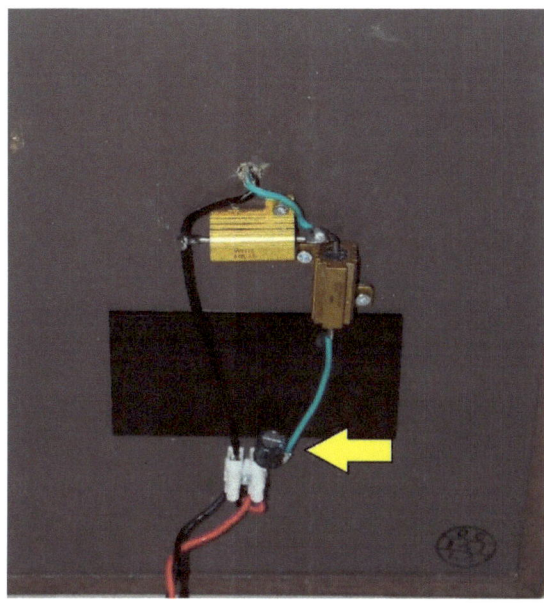

Here the two resistors that form the L-pad have been mounted on the rear panel of the enclosure. Note also the simple capacitor crossover (arrowed) on the terminal block.

Rotary L-pads

A rotary L-pad looks very much like a normal potentiometer. However, unlike a normal pot, it actually uses two variable resistors on the one shaft. These are wired so that a constant impedance is seen by the crossover as the knob is adjusted.

The L-pad must be selected so that its impedance matches the tweeter impedance, and the L-pad also must have sufficient power handling.

Using an L-pad, especially when initially setting up tweeter levels, is the easiest and most convenient approach to take. However, L-pads are often very expensive – so if you find a tweeter level control on a salvaged speaker, you should always grab it!

Note that once tweeter levels are set correctly, the L-pad can be replaced by fixed value resistors that perform the same function – but don't have such easy adjustability.

An L-pad maintains a constant impedance, even as the level is changed. This is the easiest way of setting tweeter levels, especially when initially configuring the system.

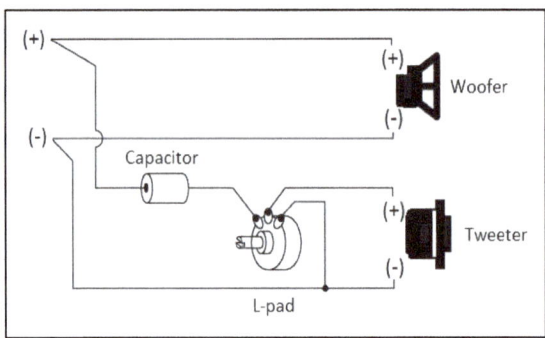

This shows the wiring of an L-pad controlling tweeter level. In this case, a simple capacitor is being used as the crossover.

Fixed resistors

Using fixed value resistors wired in an L-pad configuration is the cheapest approach to take when reducing tweeter output. Doing this is now much easier than in the past, because there are a number of free on-line calculators that can provide you with the resistor values to use.

Do a search under "L pad calculator". One available at the time of writing can be found at: http://www.diyaudioandvideo.com/Calculator/DriverAttenuationLPadCircuit/

For example, if you have a 4 ohm tweeter and it is much too loud, you may wish to drop its output level by 9dB. The above calculator then shows that to do this, R1 should be 2.6 ohms and R2 2.2 ohms.

Easiest in this situation is to buy ten or twenty 1 ohm, 5 watt resistors. Here, using the above calculator results, you'd make R1 equal 3 ohms (ie three 1-ohm resistors in series) and R2 equal 2 ohms (ie two 1-ohm resistors in series). If this sounded pretty good, you might then try to better achieve the exact values of suggested resistors (eg by series and parallel combinations).

(Note: be careful not to put R2 across the crossover rather than across the tweeter – that is, R2 is on the tweeter side of the series resistor.)

If the tweeter were 8 ohms and you wanted a similar 9dB drop, R1 = 5.2 ohms and R2 = 4.4 ohms.

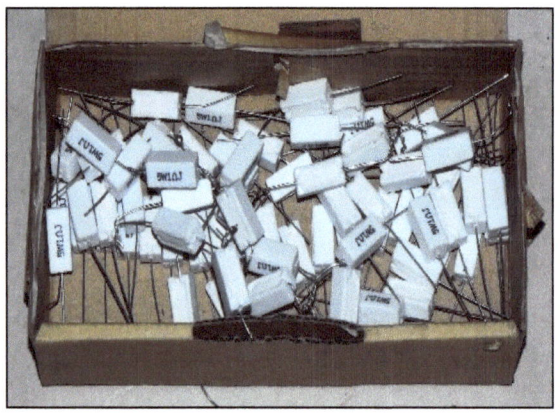

A box of 1-ohm, 5 watt resistors allows you to quickly and easily make up different values in full and half-ohm amounts.

The above calculator also displays the required wattage of the resistors. Remember, tweeter power levels are generally very low when compared with system power, so these resistors will usually be 5 or 10 watt designs.

That said, always check that these resistors are not getting too hot when the system is working hard.

Setting the levels

The electronics of reducing tweeter outputs can be cheaply and easily achieved. But how do you select the required level?

If you have a speaker that you are using as your reference, use it as your treble benchmark. Play white noise, switching back and forth between the developmental and reference speakers. (Note: adjust the volume so that the loudness of each speaker is similar.) Small changes in tweeter level can be easily picked in this way.

If you don't have a reference speaker, listen to lots of music, adjusting the tweeter level until you achieve the most pleasing result. In some respects, this is easier said than done: you'll quickly find that the amount of treble the sound engineer has included in the mix varies dramatically across different artists and songs. Therefore, a tweeter level that is perfect in one song may sound over-bright in another – or too subdued!

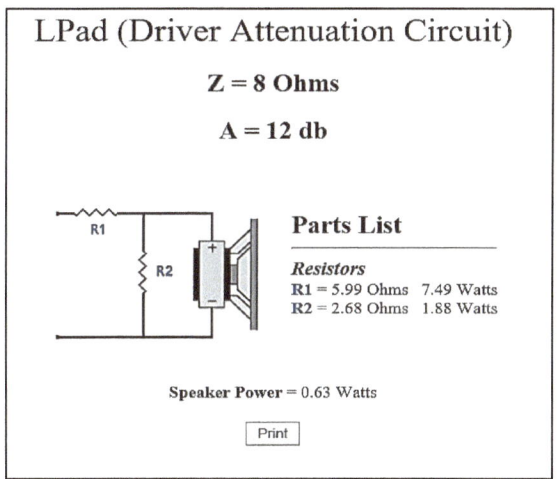

A screen grab of the L-pad on-line calculator cited on the previous page. Calculators like this one allow you to quickly and easily determine appropriate resistor values.

Woofer Tester impedance test

Woofer Tester can be used to measure the impedance curve of the complete system. Especially if you're using a crossover with unknown characteristics, testing the impedance ensures you haven't ended up with a combination of components that gives a dangerously low impedance at any point across the frequency range.

Typically, you do not want to see impedance anywhere that's much lower than the nominal system impedance (eg 4 or 8 ohms).

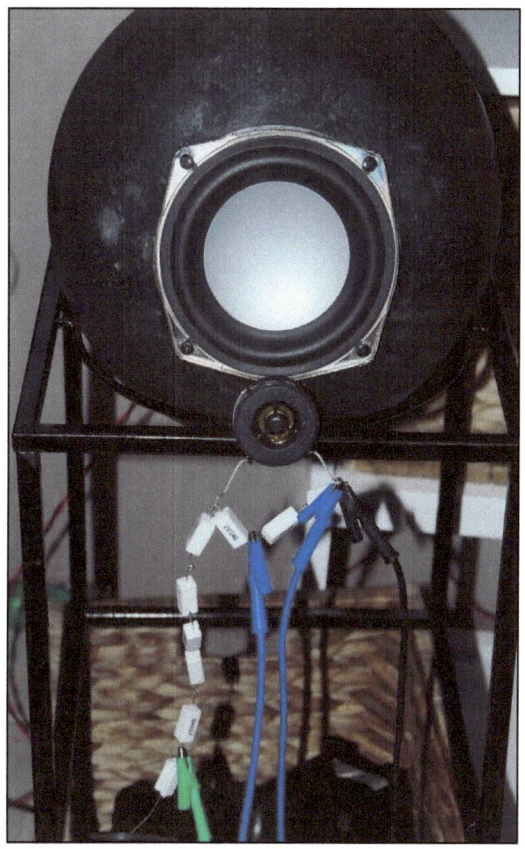

Setting tweeter levels. Below the spherical enclosure and flat-face woofer is the tweeter. Below that are two strings of 1-ohm, 5-watt resistors. One string is in parallel with the tweeter (R2 in the diagram on page 53) and the other is in series with the feed (R1 in that diagram). The crocodile clip cables allow the easy selection of different values. Once the required tweeter level was gained, a check of system impedance was carried out using Woofer Tester. More information on this speaker system starts at page 60.

Using prebuilt enclosures to build a subwoofer

- Using two pre-built car subwoofer boxes to make a custom, dual driver design with ultra-long ports

This subwoofer uses two pre-built enclosures, joined together to provide a larger volume, and also to provide the room for long ports. It's an approach that is quick and easy and can give outstanding results.

This dual-driver subwoofer was made by using two pre-built boxes joined together, forming a single acoustic enclosure. The resulting box is very stiff and allows the use of long ports. For its size and cost, this is one of the best subwoofers I've ever built.

One of the prebuilt enclosure, designed to suit a 10-inch woofer. While a ported final design is used in the final design, it is better to start with a sealed enclosure as then the port diameters can be sized correctly.

Apart from the low cost of the two boxes, taking this approach has a number of advantages over building something from scratch:

(1) The panels are all small in area and so the stiffness of the finished enclosure is quite high.

(2) Very little woodworking needs to be done and any that is required isn't critical in nature.

(3) The long, thin design is very suitable for car and home use.

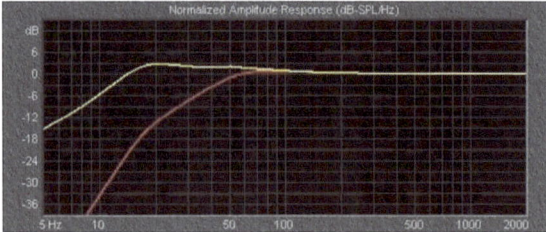

The response of the subwoofer, as predicted by the BassBox speaker design software program. The yellow line shows the response for an in-car environment, while the red line shows the modelled response within a room. In a car, there is good bass down to 20Hz!

Prebuilt boxes

When building a car subwoofer (or a house subwoofer that's not going to be in plain view), it's usually both cheaper and easier to buy prebuilt enclosures. In short, you can buy carpeted enclosures cheaper than you could ever make them yourself. However, buying a prebuilt box does not guarantee good sound – that's all in the design...

Here I am using two boxes, each with an internal volume of 23 litres. The boxes were supplied fully carpeted, with the speaker hole pre-cut and speaker terminals fitted.

Drivers

The drivers used were 10-inch units, each with 125 watts RMS power handling.

Their specs were:

Impedance:	4 ohm
Sensitivity:	88dB (1W @ 1m)
Res frequency:	33Hz
Qts:	0.502
Vas:	29.6 litres

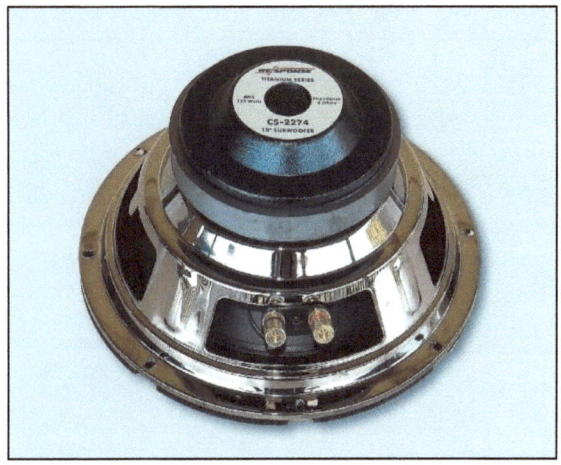

The drivers that were used were 10 inches in diameter and had good specifications. Similar car subwoofer drivers are readily available.

The bass response was optimised by tuning the enclosure to 26.5Hz using two 600mm long, 63mm diameter internal diameter ports. Modelled using BassBox software, this combination of tuned box frequency, the (actual) 37-litre total box volume and the 10-inch drivers, gave an in-car frequency response that was quite strong down to 20Hz.

Inside a home, the modelled bass response rolled off by 3dB at 40Hz. However, there was sufficient cone excursion left that this could be boosted to give bass that was audible down to 30Hz.

The ports used were commercially available flared designs, which could be easily adjusted in length by adding 65mm-diameter plastic pipe. The flared section was used at both ends of each port, reducing the chance of port noise that could otherwise occur as air flowed around the sharp inner edges.

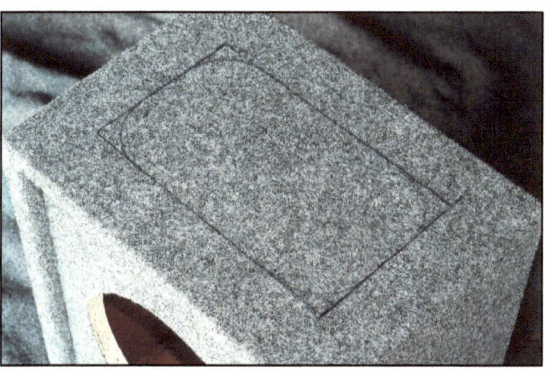

The enclosures were first modified by cutting a panel out of the end of each.

Once the opening had been cut out, the two enclosures were aligned and four holes drilled for the attachment bolts. That done, a sharp knife was used to cut away a piece of carpet all around the opening.

Lots of glue was then applied around the faces and the bolts were tightened. The excess glue was spread around the join with a wet finger. This gives an acoustic seal and also provides a stiffening flange to the enclosure.

The flared plastic vents were connected together using cheap 65mm-diameter plastic pipe. This approach makes it easy to construct ports of the required size and flow characteristics

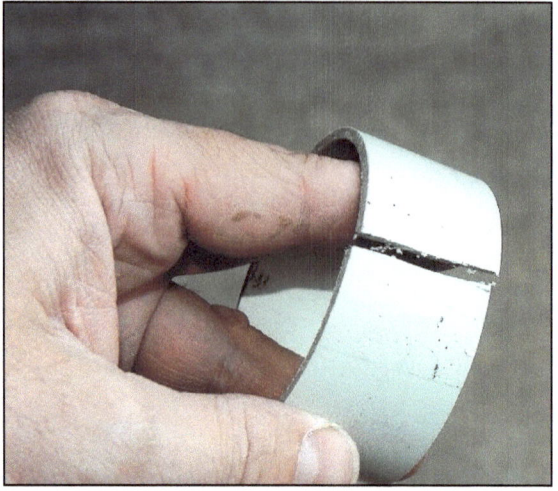

Gluing the long ports into place was not sufficient to secure them – they also required brackets to hold them rigidly inside the box. An effective bracket can be easily made by first cutting off a 30mm surplus length of the 65mm-diameter plastic pipe. That done, square the ends and then make a single cut longitudinally along the section.

This view clearly shows how the two boxes become one, and the long ports that could be fitted within the enclosure.

Next, using a heat gun, soften the pipe to one side of the cut and then bend that section outwards (use oven mitts as the pipe is hot!). With a bit more heat gun work, you should end up with a bracket which wraps itself at least halfway around the port. The bracket is attached to the inside of the enclosure using short self-tapping screws.

When extending pre-formed ports, it is a good idea to file away the lip that would otherwise occur where the pre-formed port joins the plastic pipe.

A food can makes a convenient template to mark the correct size hole for the port. Cut out the hole with a jigaw.

The enclosure was lined with quilt wadding. Ensure that the wadding does not block the ports. Glue the wadding into place so that it cannot later move.

The two drivers can be run in a variety of configurations – as a stereo pair, or in parallel and driven from a single channel amplifier. In the latter case, ensure the amp is happy driving the resulting total impedance.

Connecting the drivers

The two woofers can be driven in parallel from one amplifier or they can be driven separately by a stereo power amplifier - but there are a number of traps here.

If you want to drive the woofers in parallel, they will constitute a 2 ohm load. Many car subwoofer amplifiers will happily drive a 2 ohm load, so that is one option.

A second option is to use the two channels of a stereo amplifier to drive each woofer separately (4 ohm loads). Or, if you have a 4-channel car amplifier which can be bridged to drive 4 ohm loads, then you can use that to again drive each woofer separately.

What you must not do is connect a stereo amplifier in bridge mode to drive the two 4 ohm woofers in parallel; ie, a 2 ohm load. In this case, the separate "bridged" amplifiers will each "see" a 1 ohm load - most amplifiers cannot drive a 1 ohm load and will blow fuses or be seriously damaged.

When running the subwoofer in 2 ohm mode, simply connect the two drivers in parallel. To do this, connect the power amplifier to one set of the speaker terminals and then run more cables to the other speaker terminals, making sure that you connect positive to positive and negative to negative.

Developing a compact speaker system

> - Step-by-step process of developing a low cost compact speaker system
> - Mixture of salvaged and new parts
> - How to gain high quality sound, especially from small enclosures

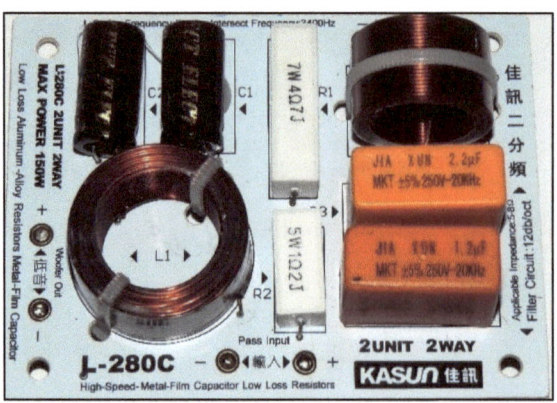

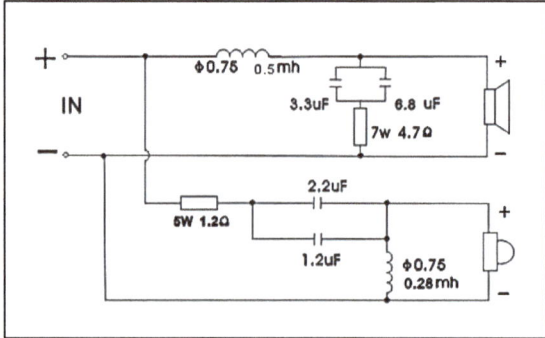

The appearance and circuit of the selected crossover. It has a nominal crossover frequency of 3400Hz and uses air-cored inductors.

Here is the step-by-step process of developing a compact speaker system, using almost all the techniques that have been covered in this book. The result is a 5-inch, two-way speaker system that sounds excellent, especially in relatively small rooms like a home office or bedroom.

The components

Both the tweeters and woofers were salvaged from second-hand systems. The woofers, that use a near flat face, are the 5-inch units salvaged from a JVC compact hi-fi speaker system. The dome tweeters were salvaged from two-way in-ceiling PA speakers.

The dome tweeters were salvaged from two-way ceiling speakers.

The tweeters use built-in capacitor crossovers - but they were bypassed in this application. Instead, a pair of new, eBay-sourced Kasun L-280C crossovers was used. These have a nominal crossover frequency of 3400Hz and a slope of 12dB / octave.

In addition to these components, variable L-pad controls are used to fine-tune tweeter levels. To give this control a wide range of adjustment, additional wire-wound resistors are used (more on this in a moment).

The L-pads (one for each speaker) were salvaged from some large old speaker enclosures that had damaged drivers.

These are the approximate costs of the components:

Woofers x 2	-	$10
Tweeters x 2	-	$14
Crossovers x 2	-	$35
L-pads x 2	-	$5
Wire-wound resistors		$5

This gives a total of $69.

Note that it is not intended that you duplicate this list by finding each of these components; it's to show that it is quite possible to obtain good quality parts at a low price – if you're prepared to look around.

The L-pads (complete with knobs) were salvaged from discarded speakers.

The enclosures

The enclosures are spherical in shape. They are formed from two 28cm diameter bowl-shaped flower pots, made from a clay-fibre mix. (These are smaller versions of the pots used in the speaker build on page 37.)

The spherical enclosures are stiff, light and easily put together. When mounted on stands a little forward from walls, they have a transparency and 'forwardness' of sound that is unusual and effective.

The enclosures are ported, with an internal port exiting on the rear. The port is made from plastic pipe.

The enclosures are lined with polyester quilt wadding and the two halves of the enclosure are glued together with water clean-up Liquid Nails building adhesive.

The approximate costs of these parts are as follows:

Bowls x 4	-	$80
Incidentals	-	$20

The enclosures therefore cost about $100 for the pair.

Medium density fibreboard could have been used to form normal boxes instead of using the spherical enclosures. This would have reduced cost but would also have been more work.

Each enclosure is formed from two bowl-shaped flower pots, made from a clay-fibre mix.

The stands

As we'll see later, the stands are an important element in the design. These place the tops of the enclosures about 1.2 metres above the ground. The 3-legged stands were made from square steel tube, 15mm in size. About 6 metres of tube was used. The stands have plastic square plugs in the tube ends (12 are needed) and have been painted black.

The approximate costs of these parts are as follows:

Tube	-	$10
Paint	-	$10
Caps	-	$5

The stands therefore cost about $25 for the pair.

Total cost

Total costs are therefore:

Components	-	$69
Enclosures	-	$100

Stands   - $25

That adds up to $194 – call it $200 in case I've forgotten something.

The design

The Thiele-Small specifications of the woofers were measured first, using Woofer Tester 2.

The specs were:

Impedance:	4 ohms
Sensitivity:	83.9dB (1W @ 1m)
Res frequency:	58Hz
Qts:	0.556
Vas:	5.4 litres

(Note the low sensitivity – the trade-off in getting a low resonant frequency from such a small driver. This low sensitivity later had implications for tweeter matching.)

Also using Woofer Tester, a suitable enclosure was modelled. The selected enclosures each had a total internal volume of about 8 litres, so that was obviously the desired modelled volume.

Using this 8 litre volume, and tuning the enclosure to 51Hz by the use of a 130mm long port 37mm in diameter, gave a -3dB point of about 45Hz. That low frequency response is good indeed for such small speakers.

Adding in room compensation (estimated as a 12dB per octave increase from 38Hz) gives a smooth increase in bass, peaking at +5dB at 53Hz.

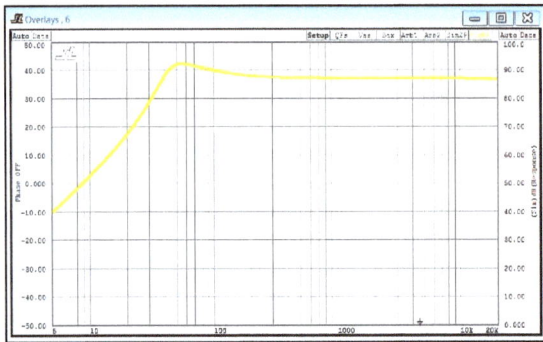

Predicted in-room response shows a smooth +5dB peak at 53Hz.

Building the enclosures

It was always intended that the tweeters would be mounted separately (ie not within the spheres) so the woofers were installed within the enclosures and the enclosures then assembled.

This process was the same as was shown for the larger spherical speakers previously covered (see page 37).

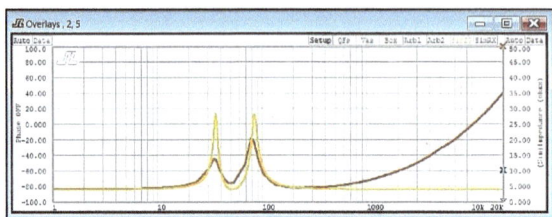

With the woofers installed in their enclosures, the speakers were tested using the 'box test' function in Woofer Tester. The yellow line shows the predicted impedance curve and the orange line the actual impedance curve. The strong correlation, especially around enclosure resonance (the double hump), can be seen.

Testing the enclosures

With the woofer installed in the ported enclosure, testing was undertaken.

Firstly, Woofer Tester was used to compare the speaker's actual impedance and phase curves with that predicted by the software. This test showed strong correlation between the predicted and actual results – that's normally the case with this excellent product.

Secondly, the enclosure was tested as a small subwoofer. Working with two small satellite speakers in a home office, the sphere proved to give excellent bass augmentation. The system used a 2.1 amplifier (2 x 50W and the sub output 1 x 100W). With just low frequency response desired, the sphere worked best close to the floor.

I also tested the sphere as a small car subwoofer. If huge bass wasn't wanted (instead just an added low frequency presence of the sort achieved by many original equipment wagon and hatch subs), it again worked well.

However, while it was good in a 'small sub' application, that wasn't the enclosure's intended purpose!

The enclosure being tested as a small domestic subwoofer. Here it's been placed in a stand sold for garden pots, and equipped with a grille borrowed from a salvaged Philips speaker. In a bedroom or home office, and when matched with small satellite speakers as a 2.1 system, the little sub worked well.

The speaker being tested as a small car subwoofer. If earth-shattering bass was not required (instead just a lift in low frequencies), it worked well.

I then wired into place the tweeters, just placing them at this stage near to the spheres. Initially, I used the tweeter's inbuilt capacitor as the crossover, but the result was poor.

I then bypassed the capacitor by soldering a wire in parallel with it, and wired-in the commercial crossovers. (Actually, I first started using some Alpine crossovers that I'd previously bought. But to my ears they rolled off the bass response a bit and caused the treble to be peaky.)

With the eBay crossovers installed, the sound was much better – but the treble was overly strong. Using an on-line calculator to determine the values, I installed wire-wound (ie high power) resistors wired in an L-pad configuration to drop the tweeter output. Woofer Tester measured the tweeter's impedance at 8 ohms, and according to the online calculator, as much as 14dB needed to be pulled from the tweeter's output before the treble became balanced. (However, whether it was or was not actually 14dB is problematic: the calculator's values depend on the actual speaker impedance at those frequencies.)

With the tweeter levels set in this way, I was fairly happy with the sound. However, it quickly became obvious that relatively small changes in the height of the spheres off the floor, their proximity to walls, and the exact tweeter placement all made major changes to the way the system sounded.

In fact, there was no point at this stage in setting the exact tweeter levels, when these other changes made so much difference. It was time to do some testing of placement!

I initially had the tweeters sitting slightly below, and at much the same 'forwardness', as the woofers. I then attached the tweeter to a long timber stick, and with the help of an assistant, had the tweeter moved around as I listened. (And then I did the tweeter moving as my assistant listened.)

This showed that tweeter placement made a major difference to the sound. Best results, in our judgement, came when the tweeter was placed forward of the woofer face by about 95mm. (Note that this is in sharp contrast with many speaker approaches where the tweeter is stepped backwards from the woofer.) Furthermore, the tweeter sounded best when it was raised so that it's vertical level roughly coincided with the lower part of the woofer roll surround.

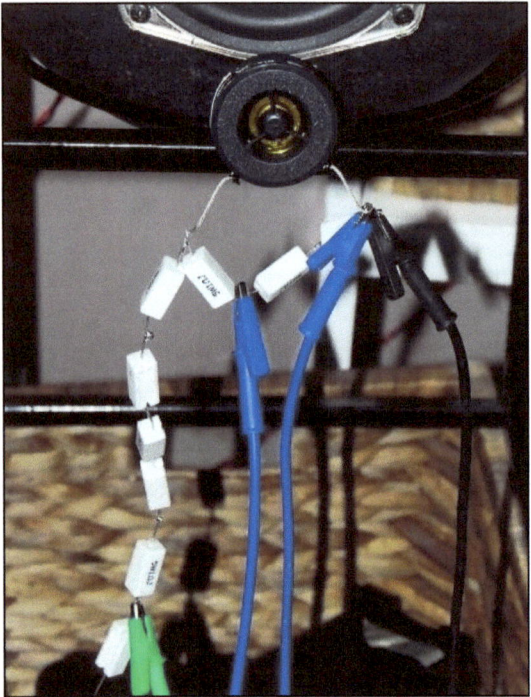

Using 1-ohm wire-wound resistors to initially set the tweeter level. Here the tweeter is being held to the metal frame by just its magnet.

The differences to the sound made in moving the tweeter around were clear and pronounced; I imagine a similar outcome would occur when working with conventional shaped speaker enclosures as well – so I suggest that you don't cut a hole in the baffle for the tweeter until doing this experiment.

Next, variations were trialled in the height from the ground of the spherical enclosure / tweeter combination. This showed that the cleanest bass occurred when the speaker was about a metre off the ground. (Too close to the ground and the bass became muddy – less a problem with a dedicated subwoofer.)

In terms of proximity to walls, best result came with the speaker about 20-40cm forward of the wall. Note that this testing was carried out in my home office – a relatively small room about 4 metres x 3 metres, with about one-third of the wall space covered in open shelving carrying an assortment of stuff (ie non-reflecting of high frequencies).

With the height of the main enclosure and the location of the tweeter finalised, the stands could be made.

I am lucky enough to have an extensively equipped home workshop so I was able to bend the steel tube in a hand tube bender and use a MIG welder to join the legs.

The stands each use three pieces; the rear two pieces forming legs splayed at the top and bottom. The upper 'splay' carries the sphere; the lower splay gives the stand lateral stability. The third leg is the front one. It is shaped differently to the other legs, with a tweeter mount protruding at the front and the lower leg further forward than the spacing between the rear legs. With the heavier driver mounted at the front, this provides good stability.

With the spheres mounted on the stands, and the tweeters attached to the forward projection of the stand, I could so some serious listening.

I then decided that I needed better tuning of the tweeter level, so I removed the resistor strings and wired-in the rotary L-pads. However, and not surprisingly, these needed to be set at about '1' on a scale of '10' – so fine adjustment was difficult.

I then provided about 10dB of tweeter reduction via fixed resistors, run in addition to the L-pads. This gave the adjustable L-pads quite fine control.

So how do the speakers sound? I don't want to be one of those people who says every speaker design in their books is fantastic - *in every single respect!*

So if I were to be ultra-picky, I'd say the treble can be a fraction peaky (there is sibilance with some vocalists) and the midrange is a little lacking – specific vocalists can sometimes be more lost in the mix than they really should be.

But having said that, the bass is excellent, the imaging superb and the general quality of sound emanating from these tiny enclosures amazingly good.

Worth adding mid-ranges and new 3-way crossovers? Probably – but you've got to stop somewhere!

www.ingramcontent.com/pod-product-compliance
Lightning Source LLC
Chambersburg PA
CBHW050749180526
45159CB00003B/1394